Blossoms

And the Genes That Make Them

MAXINE F. SINGER

OXFORD
UNIVERSITY PRESS

OXFORD
UNIVERSITY PRESS

Great Clarendon Street, Oxford, OX2 6DP,
United Kingdom

Oxford University Press is a department of the University of Oxford.
It furthers the University's objective of excellence in research, scholarship,
and education by publishing worldwide. Oxford is a registered trade mark of
Oxford University Press in the UK and in certain other countries

© Maxine F. Singer 2018

The moral rights of the author have been asserted

First Edition published in 2018
Impression: 1

Published in the United States of America by Oxford University Press
198 Madison Avenue, New York, NY 10016, United States of America

British Library Cataloguing in Publication Data
Data available

Library of Congress Control Number: 2017947207

ISBN 978–0–19–881113–8

Printed in Great Britain by
Clays Ltd, St Ives plc

For Luna, Kapp, Elliot, and Emma,
and the gardens they will cherish.

ACKNOWLEDGEMENTS

The stories in this book are based on the published scientific work of many extraordinary scientists. I have not associated their names with the discoveries chronicled in this book. According to contemporary publishing habits in biology, the name of the senior person in the laboratory is usually last in the list of authors. Often, the work is, nevertheless, identified by and credited to that person's name. Forgotten in this method are the several students and postdoctoral fellows who actually did the laboratory and greenhouse work under the direction and inspiration of the senior scientist. That senior is also responsible for obtaining the financial resources required for the work. I am indebted to the published scientific papers of all these scientists for teaching me about flowering and making it possible to share their stories with readers.

Elliot Meyerowitz at Caltech is a leading plant biologist whose work has had a major influence on scientific descriptions of flowering. His generosity with ideas and clarifications has been a great help to me as I educated myself into the marvelous stories about blossoming. I am very grateful for his advice; it helped me clarify some complex mechanisms. I thank Laura Youman Debole for the lovely line drawings that grace this book.

I also thank my colleague Susanne Garvey at the Carnegie Institution for Science for reading the whole manuscript and providing suggestions. She is one of those rare humanists who knows how to read science.

Any errors that remain in the text are my own doing.

Finally, I thank my friend of many decades, Naomi Felsenfeld, for being a great companion as a knowledgeable wild flower hunter. She does insist on one very helpful caution, namely, not to bother trying to identify the many small, daisy like flowers. She says the same thing about small brown birds…but birds are her thing, not mine.

CONTENTS

CONTENTS

PART IV
Shaping a Flower

PART V
Decorating a Flower

INTRODUCTION

Sometimes, again, you see them occupied for hours together in spoiling a pretty flower with pointed instruments, out of a stupid curiosity to know what the flower is made of. Is its colour any prettier or its scent any sweeter, when you know?

Wilkie Collins, *The Moonstone* (1868)

Who doesn't love flowers? Bright blossoms enhance the everyday lives of many all over the world. Even the dirt yards around shacks in the poorest slums often have an old oil barrel splendid with plants in bloom. The first things many children draw are brightly colored flowers with a yellow sun shining down upon them; somehow, they realize that light is important. Artists never outgrow this desire to illustrate nature. For centuries they have painted wild flowers and masses of bright blooms in vases. But how many of us have wondered how a real plant makes a flower?

Over the past few decades, scientists have been working on just that. Thanks to modern methods of genetics and biochemistry, and the use of a tiny weed called thale cress or mouse cress, or, more formally, *Arabidopsis thaliana*, we can now tell the remarkable story of when and how a plant makes a flower. Botanists have answered Wilkie Collins' complaint with proof that it's not just Feynman's Jupiter that is wondrous.

This story could not have been written thirty years ago. Then, no one knew how and when a plant begins to make a flower. No one knew how, at an appropriate time, plants decorate themselves with colorful and perfumed blossoms.

Flowers did not evolve for our pleasure. Their purpose is to ensure the production of seeds. The more seeds, the more new plants can be made, as long as the environment is hospitable and provides light, water, and nutrients. They evolved to optimize seed production by attracting insects or birds through the color, fragrance, and form of their flowers. These visiting animals pick up pollen and then spread the pollen so it can fertilize a plant's egg that then forms seeds. The egg can be in the same flower or in another flower of the same species. Some plants have no need of pollinators; they spread their pollen on the winds.

Flowers matter to us for much more than their beauty. They are the source of food for humans and animals. The staples of human diets, worldwide, are seeds including rice, corn, wheat, and beans. The fruits that carry seeds are another major food source: fruits such as tomatoes, squash, apples, oranges, and so many others.

Flowers are also big business. That industry goes back at least to Roman times, and is behind every flower we see in a flower shop and every plant that fills nurseries. As long ago as the sixteenth century, tulips imported from the Ottoman Empire were the buzz of Europe. By the middle of the next century, Holland was consumed by tulipomania. Speculators dealt in tulip bulbs, not stock. Frenzied collectors mortgaged houses, estates, even their horses to pay for tulips long before the bulbs were actually available. But, like a stock market, the tulip market eventually crashed. The tulip stock had been destroyed by a virus.

Today, flowers are a multibillion dollar global industry. In the USA in 2005, people spent, on average, almost $26 a person on flowers, a lot less than in most European countries. The Swiss are

the biggest spenders, at almost four times US outlays. It's no wonder that flower growers spend considerable sums to ship fresh
flowers worldwide by air; they know there is a good market. Most
of the cut flowers in US flower shops come from Latin America.
Big flower shows, like the annual one in Philadelphia, can bring
billions of tourist dollars to a city. The science behind flowering
matters for agriculture and industry too.

* * *

In the middle of the nineteenth century, in the city of Brno, in what
is now the Czech Republic, the monk Gregor Mendel experimented
with plants. He learned that seeds contained information, what he
called "factors," for making a plant as well as determining the color
of the flowers and the shape of the seeds. Today, we call those factors genes. And we know that genes determine not only the color
of the flowers but their shapes, their various parts, their perfumes,
and even when a plant will form flowers. We also know that genes, be
they in plants or animals, bacteria or viruses, are parts of DNA, that
beautiful double helical molecule that became an icon of the
twentieth century. Questions about when and how a plant makes
flowers found answers in genes and DNA. This book tells the story
of what we know so far about how it all works.

It's too bad that most people learn about genes from the media,
which usually report only problematic genes that cause human
disease or the imagined dangers associated with genetically modified plants. That gives genes a bad name. They deserve a better
press. After all, they are also responsible for the beauty of roses
and the scent of jasmine.

Flowers appeared quite late in the evolution of life on planet
Earth. According to the fossil record, land plants themselves

appeared only about 500 million years ago. The earliest flower fossils date to about 130 million years ago. But because flowers are fragile, and likely to decay before fossils can form, they may have evolved a little earlier but not been preserved. Still, flowering plants have been amazingly successful; they occupy all sorts of habitats on Earth, from the frigid polar regions to the tropics, from lowlands to mountain tops. And their diversity of color, shape, and perfume is dazzling.

Charles Darwin, knowing nothing of genes and how they drive evolution, found only frustration in the diversity of flowers. He described the history of flowers on Earth as "an abominable mystery." He might be pleased to learn that a lot of the mystery has been cleared up by new fossil finds, and especially by the study of DNA. It seems likely that the ancestors of flowering plants were ferns. The flowering bush called *Amborella*, which grows on a South Pacific Island, seems to be closely related to the most ancient ancestor of existing flowering plants. Water lilies are the most ancient kind of plant most of us are likely to see.

Once begun, the evolution of flowering plants followed the same kind of scenario as the evolution of animals. Genes in flowering plants are constantly turning up new varieties, because mutations arise in their DNA as a result of radiation, exposure to particular chemicals, or mistakes made in copying the DNA when cells divide. If those changes result in traits that prove to be an advantage for the plant in its particular environment or that in which its seeds land, the altered genes will be passed on to new generations of plants. If the mutations make the plant less fit for its environment, the plant may die without leaving seeds carrying the altered genes. In this way, new varieties and, eventually, new species that are well adapted to their environments arise and flourish—at least while that environment lasts. That, in short, is natural selection. Over more than 100 million years, this process, together with changing climates and evolving animals, have resulted in the huge variety of flowering plants on Earth.

All the elements of a flower are the consequence of the genes packaged into the plant's seeds. Like animal sex, the scheme is circular: genes build the plant and its flowers so that those same genes are preserved in seeds that can make more plants and flowers. Flowers are the sex organs of plants.

An artist pondering painting a flower may consider what kind of flower it might be, what colors, what shape and size, how many petals? The choices available to an artist are restricted by his imagination and the materials available. The choices available to a plant are restricted by its genes.

Being a city person, I was entering middle age before I came to appreciate gardens and flowers. It was even later, after a few summers in the mountains of Idaho, before I paid serious attention to wild flowers. But I knew about genes and DNA long before that. They have been at the center of more than 50 years of my own scientific research, although that work concerned the genes of bacteria, viruses, and animals, not plants.

Long before anyone knew about genes, they knew that plants sometimes grew with unexpected characteristics. For millennia, farmers and gardeners took advantage of the new properties to improve food crops and the appeal and diversity of cultivated flowers. They learned that seeds taken from the unusual plants would grow into similarly unusual offspring. They selected and saved those seeds. Such artificial selection was a model for Darwin's proposal of natural selection as the mechanism for evolution in nature. All this made a lot more sense when Mendel discovered genes. Although they were contemporaries in time, Darwin never learned about Mendel's experiments. That was really too bad, because genes solved one of Darwin's conundrums: what causes the variety upon which natural selection works?

One way in which the genes in seeds may be different from those in the plant that made them is if the plant was pollinated by pollen from another, related plant, rather than from its own pollen. If the two plants are entirely different species, they are not likely to be cross-pollinated at all. But chance mutations in the genes of two plants of the same species give rise to plants whose genes are slightly different from those in either parent. Even if the plant self-pollinates, the genes in the egg and those in the pollen can differ somewhat from one another. In this way, the seeds from pink-flowered plants may, for example, produce new plants with white flowers or even white flowers with pink spots. Evolution takes advantage of these changes to produce diversity.

Part I of this book introduces some of the ideas that underlie the science that follows. Stories about what genes are and how they work, derived from research, form Part II. Part III describes how a plant knows when it is time to flower. Part IV is concerned with how plants construct flowers, while Part V summarizes the ways in which a plant decorates its flowers so as to attract pollinators. Readers will encounter some new concepts and unfamiliar terms, because that is the only way we can talk about new ideas or newly discovered genes and processes. The key new words that will crop up from time to time are in bold print; their definitions are in the glossary.

The steps and processes behind the making of a flower are not always logical and some may seem unnecessarily complex. This is because evolution is undirected, haphazard, and often inefficient; it can take advantage of quirky changes in genes and how those

changes fare in a particular environment depending on temperature, light abundance, and the availability of essential insect or
bird pollinators. As the late biologist and writer Stephen Jay Gould
always emphasized, evolution is *contingent*—contingent on a particular gene change in a particular environment.

Absent from the text are the names of the dozens of scientists
who did the experiments and developed the stories. They are left
out for several reasons. Often, there is a long history of discovery
and development leading to knowledge of any particular process
or gene. It's not always simple to sort out just who should be mentioned in connection with particular findings. Several different scientists may make discoveries, sometimes over decades, that only
come together into a coherent story years later. The story of when
and how a flower is made is complicated enough without adding
the complexity of the history of discovery. For those who are interested, the list of suggested further reading at the end of the book
can be used to find out more about the many scientists who have
contributed to the story.

Our understanding of how and when plants flower is far from
complete. There are many aspects that still need to be explored
and explained. And some of what is described here, while based
on current science, will surely turn out to be at best misleading
and, at worst, wrong. That's the nature of science. New experiments and concepts aim to bring us closer and closer to the way
things really work.

My hope is that after reading this book, whether you see long-
stemmed red roses in a street kiosk, or yellow tulips in spring
gardens, or wild purple asters covering a mountainside in late
summer, you will remember that they are the result of genes
working and evolving over the long history of life on Earth.

PART I

What Plants Are and What They Can Do

Mary Mary quite contrary,
How does your garden grow?
With silver bells and cockle shells
And pretty maids all in a row.

English Nursery Rhyme

They may be stuck in one place, but plants can do many things. We will look at some of them in this section. Along the way, it introduces some of the words and concepts needed to tell the stories that follow.

Chapter 1 is about names: the names of plants, of flower parts, and of genes.

Vocabulary is the bugaboo of many efforts to write about science, and genetics in particular, for the wider readership. Nobody seems very bothered by encountering complex names from an unfamiliar language when it comes to a novel. New words in books about science seem to present many more difficulties than names in *War and Peace*. Technical jargon turns away a lot of people. Years ago, a colleague of mine discovered that students in an introductory college biology class had to learn more new words than those in a first year French course! No wonder so many students are put off biology. Newly discovered things or processes require new names—like the verb "to google." Without them, we would be tongue-tied.

Readers will encounter a number of unfamiliar terms in this book, but I have made a real effort to keep them to a minimum. Still, all the plants and plant parts, as well as the genes, and molecules that shape them need to be called something, and many of the names used in this book were coined in the past few decades. Plant names were a challenge even in the past. These days, the same gene is given different names if it's discovered in more than one laboratory or more than one organism. This is frustrating for scientists reading research papers. Scientific societies try to ameliorate the problem by establishing committees where decisions on uniform names are fought out, but the problem has not gotten much easier over the centuries.

Chapter 2 is about the similarities and differences between plants and animals. The differences are obvious to everyone, while the similarities may be a surprise. Our kinship with plants goes deep. At that very basic level, we are cousins: distant cousins, but still cousins. How distant? No one knows for sure but the link, what is called our "last common ancestor," could have lived as far back as a billion years ago, when the ancestors of land plants appeared in the sea.

Plants know a lot about their environments. They may not have our sense organs, but they respond to the length of the day, to the direction of sunlight, to temperature and the pull of gravity, even to touch in some cases. They also sense their age, their level of maturity. Chapter 3 summarizes how plants know some of these things even without the senses that inform animals. As we shall see later, all these aspects matter for flowering.

Chapter 1
NAMES

Theophrastus was the first person known to try to bring some order into the naming of plants. Consequently, he is sometimes called the "father of botany." Living around 300 BC, he was lucky enough to become a student of Aristotle. A monument to Theophrastus stands in his birthplace, the small town of Eressos on the Greek island of Lesvos. Eressos must have been a special place; some three centuries before Theophrastus, the great poet Sappho was born there. Theophrastus' groupings were, by modern standards, not very useful, but they were a start: trees, shrubs, herbs, cereals, and the presence or absence of thorns. For whatever reason, though, he had little to say about flowers.

Eighteen centuries later, in Europe, herbalists and botanists began to invent and promote standard names to avoid the confusion caused by the different names used in different places for the same plant. They were forced to do this because, after Columbus's voyage to America in 1492 and the dawn of the so-called Age of Exploration, many new plants were being imported from the western hemisphere. Botanical gardens were established at this time to collect common and uncommon plants.

The naming problem was finally resolved in the eighteenth century by Linnaeus, a Swedish botanist. Linnaeus adopted some earlier ideas, including the use of Latin, then the language of most scholars and physicians, and corresponded with scientists around the world as he developed his naming system. Each plant (and animal) would be known by two names. The first described the general group to which the plant belonged, the **genus** (plural, genera). The second name designated the particular plant, the **species**. Local, common names would still be used, but professional gardeners and scientists were able to communicate across geographic boundaries without confusion. And the Linnaean system is used to this day. Both names are written in italics, and the species name, by convention, begins with a lower case letter.

The lungwort that is just showing its spotted leaves in my garden as I write this in mid-March is called *Pulmonaria officinalis*. The common name, lungwort, is not very pretty, but the plant, with its early pink-purple blooms, is a favorite of mine. The dandelions that will soon show themselves are *Tavaxacum officinale*. The two plants have similar species names, but belong to different genera, and are unrelated.

This book will use common American–English names for flowers and the plants that produce them. Those common names can still be confusing. *Zea mays* is the standard name for what is called corn in the United States and maize in the United Kingdom, where the word corn is used to describe the chief cereal crop of a region, whatever it may be. Oats are called corn in Scotland, while wheat is called corn in England.

Many common flower names used in this book will be familiar to gardeners and other flower lovers. Among these are roses, petunias, and snapdragons. Much of what we know about flower color, for example, comes from studying petunias and snapdragons. One plant name that is not very familiar to most people but is in the daily vocabulary of plant scientists worldwide is **Arabidopsis**. Just as the mouse has been used as a model to study the genetics fundamental to all mammals, including ourselves, the tiny, worldwide weed *Arabidopsis thaliana* is the favorite experimental model for plant scientists. The common name of this species of *Arabidopsis* is thale cress or mouse-ear cress, but we will simply refer to it as Arabidopsis.

Most gardeners know the words annual and perennial. Seeds of annual plants are sown each year and they germinate, grow, flower, and set seed within a relatively short period. Many wild flowers are annuals. Zinnias and petunias are annuals frequently seen in temperate-zone gardens. Annual plants die back after they flower, or when the weather becomes cooler and days get shorter, leaving only their seeds to assure that they will grow again. Perennials are plants that emerge and flower year after year, such as cyclamen, forget-me-nots, and columbine. Other perennials are trees and shrubs, or vines that flower annually, including apple trees, azaleas, peonies, and honeysuckle. Some species of plants include those that are annual and others that are perennial. Arabidopsis is such a plant, and its several forms are one reason it is a good experimental organism.

Gardeners, and those who stayed awake in biology class, may recall the names of the parts of a flower. Biologists call these parts "organs," because each has a particular function, like the organs of our bodies.

Linnaeus introduced the idea that plants could be classified and named on the basis of the number of **stamens** and **carpels**, the male and female parts of a flower, respectively. He even called the stamens, which hold the pollen, "husbands" and the **carpels** "wives," and referred to the flowers themselves as "beds." Some contemporaries regarded this so-called "sexual system" with amusement, but others were shocked at what they called "botanical pornography." Some even argued that ladies shouldn't deliberately look at the sexual parts of plants. Despite this disapproval, the sexual system soon caught on because it was so accessible and straightforward. It represented a democratization of science in that anyone, not just a specialist, could identify the species of a plant by counting the number of male and female parts in the flower.

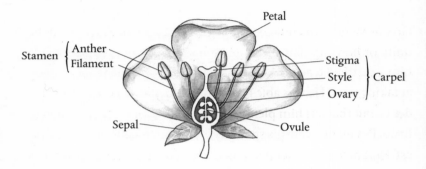

Simple flowers usually have a single carpel in dead center. It often looks like the pestle of a pestle and mortar, and is also often called a **pistil**, derived from the Latin for pestle. Carpels usually have three parts that together comprise the female reproductive organs. At the base of the carpel is the ovary that contains the egg. The ovary is topped by a tube, the **style**, at the tip of which is a knob, the **stigma**, which serves to capture pollen. Varying numbers of stamens surround the carpel. They are built of a thin filament topped by a knob, the **anther**, that carries the pollen, the equivalent of animal sperm (Plate 1).

Petals, the (commonly) colored, eye-catching part of a flower, surround the carpel and stamens. Underneath the petals, on the outside of the flower, are a few small leaf-like structures, usually green, the **sepals.** Before the flower opens, the closed-up sepals cover the flower bud. Sometimes, as in orchids and crocuses, sepals are colored, look like petals, and form part of the flower (Plate 2). The stamens and carpel, the organs that ensure the next generation of plant, are the real business of the flower. The sepals protect those organs, while the function of the petals is both to protect and to lure in pollinating birds and insects.

* * *

Gregor Mendel, the nineteenth-century monk who discovered the units of heredity, failed the examination that would have made him a certified teacher. Like many smart people, he was not good at taking tests. He was able to continue teaching the youngest students, but that left him plenty of time for his monastery garden in Brno. Pea plants were his specialty, but by no means his only interest. He, and many other plant breeders of the time, knew that seeds carried the traits of the plant that produced them—traits like seed and flower shape and color. But Mendel did something unique: he counted and kept careful records. He recorded the color and shape of the pea seeds (the peas) and the pea flowers that provided the pollen. He recorded the color of the flower on whose stigma he brushed that pollen, being careful that the flower's own pollen could not reach the stigma. Then, when the new seeds appeared, he counted how many were green and how many yellow. When the seeds were planted, he counted how many of the new plants had white flowers and how many purple, and how many of their peas were green and how many yellow. He did this over several generations, each time recording the outcomes. From these and further experiments using other traits of the pea plants, he deduced the existence and pattern of inheritance of what he called "factors" and we now call **genes**.

Mendel gave talks to interested people in Brno and even published a paper in a local journal. But few people read the paper or understood the significance of what he had done. He and his work were forgotten.

Fifty years later, early in the twentieth century, scientists doing the same kind of experiments in both plants and animals rediscovered Mendel's reports and recognized the importance of his work. This unknown monk had "scooped" them decades before they had even begun their own work, and they hadn't even known about him. After that, experiments, ideas, and new knowledge about genes came along at an ever-increasing pace.

One of the most important developments was the recognition that fast-breeding fruit flies (*Drosophila melanogaster*) were a fine model organism for studying genes. A fly gene was one of the first genes to be named. The gene's discoverer, the American Thomas Hunt Morgan, called it *WHITE* because the flies had white rather than the usual reddish-orange eyes (gene names are given in capital letters and, like Latin names of organisms, in italics).

Morgan's discovery and naming of the *WHITE* gene illustrates an odd aspect of the naming of genes. Fruit flies normally have reddish-orange eyes. If a gene required to make the reddish-orange pigment is lost or not able to function, the eyes are white. The gene's name, *WHITE*, tells us what happens when something is *wrong*, not what the gene does when it is working properly. This seems confusing and contrarian, but it does makes some sense, because for most of the twentieth century, the existence of a gene only became apparent when it did not function properly. That is still the way many genes are discovered and named. Think of the human inherited disease called cystic fibrosis. The faulty human gene that causes the disease is called the *CYSTIC FIBROSIS* gene. That does not tell us what the normal gene does, only what happens when it is not doing its job. The same upside-down situation exists in plants. *LEAFY* is a gene needed to make flowers. It was discovered when a faulty *LEAFY* gene made it impossible for an Arabidopsis plant to make flowers, so it just kept on making leaves.

LEAFY is one of the many genes it takes to make a flower and a lot of these are also named for what happens when they don't function properly. Genes named *APETALA 1* or *2* or *3* are essential for flower formation; plants with faulty *APETALA* genes cannot make proper flowers. We will come across these genes again later in the book.

Although this system for naming genes seems backward, it does at least hint at what the functional gene does. The *WHITE* gene, for example, is important for the color of a fly's eye. *APETALA* genes are necessary for making petals. A gene that is so altered that it is unable to accomplish its usual task is called a mutant gene or a **mutation**. Studying mutations has been the key to understanding genes ever since Morgan discovered *WHITE*.

Mendel figured out that the inheritance of plant and flower characteristics depends on what we now call genes. Morgan and others learned that animal inheritance also depends on genes. Breeding experimental organisms like peas and flies revealed a huge amount about genes and mutations. Early in the twentieth century, scientists figured out that humans too have genes and pass them on to their offspring. They did this by studying the way mutant traits, such as inherited diseases, are passed from parents to offspring. Amazingly, all these insights into genes and how they work were carried out without any idea of what a gene actually is or how genes cause such profound changes in living things.

Imagine then the excitement when, in the middle of the twentieth century, scientists learned what genes actually are and how they work. I was a graduate student in those years. Everything we were studying, and a lot of what was in our textbooks, took on new significance. It was a revolution. Like all revolutions, the young students embraced it more quickly than did the professors.

One consequence of the revolution was a clear definition of a gene—or so we thought. That clarity has gotten cloudier and cloudier in the sixty or more years since then. Biologists now have trouble agreeing on a straightforward definition of a gene. Definitions that seem appropriate for a while often fail in view of new discoveries. This situation may seem unsatisfactory, but it's best not to spend time worrying about it. It's much more fun to think about what we know and what we should try to learn than dwelling on definitions.

For now, it's enough to recognize that genes are units of biological information, coded by DNA, that determine how an animal or plant will grow and mature, what it will look like, how it will obtain and use energy, how it will react to its environment, and how it will reproduce. The rest of this book is about how plants and their flowers ensure that there will be future generations.

Chapter 2

PLANTS ARE LIKE ANIMALS, ONLY DIFFERENT

Most plants are green. Not many animals are green...a few insects, some frogs, some fish. Why then do we so often imagine that visitors from outer space are "little green men"? Are we jealous of the plants? We might well be. Plants manufacture their own food from sunlight, carbon dioxide (CO_2), and water. Animals can't do that; they depend on plants for food. That includes meat-eaters, because the meat comes from animals that depend on plants for their food. The process that gives plants the ability to make food for so much of life on Earth depends on that green color, and it is called **photosynthesis**. The color comes from a pigment called **chlorophyll** that is related to the red pigment hemoglobin which makes our blood cells red. Photosynthesis allows plants to grow to huge sizes. Not even the largest living animals can compete with the size of most trees.

Unlike animals, plants can't move about to find new foods or seek a mate. They have to remain where they take root. Flowers and the seeds they produce are the way plants get around this disadvantage.

Some plants fertilize their eggs with their own pollen to start seed production; maize plants are a good example of this. Others spread their pollen either by relying on the wind or with the help of pollinating insects and birds.

Even with all these differences, plants and animals are cousins, not least because they all have genes made of DNA and pass those genes to their offspring. Plants pass their DNA along in the eggs and pollen while animals endow a new generation with DNA from eggs and sperm. Microorganisms like bacteria and archaea also depend on DNA and genes, although they are not sexual organisms; they pass their DNA along when they reproduce by dividing. The dependence of all organisms on Earth on DNA and genes is profound evidence for concluding that they all evolved from a common, very ancient, ancestor that lived several billion years ago.

All life on our planet has something in common besides DNA and genes. All living organisms are made of cells. Humans are made of trillions of cells. The tiny nematode worm that many biologists study has 959 body cells. Plants have millions to billions of cells, depending on their size.

Bacteria as well as some animals and plants are single, independent cells, not easily seen without a microscope. The essentials of plant and animal reproduction—eggs, pollen, and sperm—are themselves single cells. An egg cell that has been fertilized by pollen or sperm is also a single cell. Most important, the fertilized egg contains DNA from each of its parents.

The common ancestor of plants and animals was a population of single-cell organisms. Some of these developed chlorophyll and a green color and gave rise to the ancestor of green algae and the multicellular green plants. Others, less colorful, developed into single-cell fungi and multicellular animals. The emergence of multicellularity occurred separately in plants and animals.

Cells are bags of chemicals. The bags are made of fatty membranes that do a good job of keeping the cell intact and protected. These membranes are like the security fences that surround protected army bases or industrial plants. But just as protected sites need to allow for the entry and exit of people and vehicles, cells need to allow some substances, like nutrients, to enter and other molecules, like waste, to leave. The fatty cell membranes are studded with molecules that control the intake and removal of molecules. Inside the cell, molecules large and small are organized into tiny structures; these molecules include **proteins**, carbohydrates, **DNA**, and **RNA**. Many of these structures are machines that carry out the processes needed to keep the cells alive, growing, and dividing.

Inside each animal and plant cell is another bag, called the nucleus. And inside the nucleus is the DNA, the information center for the cell. The cells of bacteria and archaea do not have nuclei; their cells are of simpler structure, and their DNA is within the cell along with a lot of other molecules.

The development of a new plant or animal begins when the fertilized egg cell divides into two, then four, and on and on, until the whole organism is constructed. At each cell division, the DNA is duplicated and each of the two new cells, the daughter cells, receives a full complement of the DNA present in the fertilized egg. The cells formed early in development are called **stem cells**, so named because they can give rise to a variety of cell types. Often, these stem cells divide to produce additional stem cells. Sometimes, however, the cell divisions instead produce two different kinds of cells: another stem cell and a cell now reprogrammed, through subtle changes in the way the genes are used, to produce a population of cells with special abilities, such as the formation of particular organs like roots, stems, leaves, or flowers. These plant stem cells are like human embryonic stem cells, except that no ethical issues arise if plant stem cells are used for experiments.

Most people know that animals make special kinds of cells—**gametes**—whose job is to produce the next generation. Eventually, the eggs produced by females and sperm from males must meet, either inside the body of the mother, as in some insects and mammals including humans, or outside in the environment, as with fish. The gametes that will give rise to eggs and sperm are set aside early in the development of an animal. In mammals, including humans, this occurs early in the development of an embryo. These cells are protected until it is time for them to mature.

Flowering plants also have gametes. But plants don't set these reproductive cells aside early in development. Instead, they form later in the plant's life, as part of making a flower. Those gametes—eggs and pollen—also come from stem cells, but they only develop when a plant has already matured enough to make a flower, at which time pollen is produced within the stigma, and an egg within the carpel of the flower.

Plant stem cells collect at the very tip of the growing shoot, which is called a **meristem**. When the meristem cells divide, one of the two new cells forms part of the plant itself—a longer stem, or the beginning of a leaf, for example. The other offspring of the divided meristem cell remains in the meristem as a stem cell to repeat the process. Then, when it is time to flower, some meristem cells receive signals that instruct them to change their program and prepare to produce cells for flower construction. It's a complicated task because the four flower organs—the carpel, stamens, petals, and sepals—will all contain different and specialized cells. Together, these four organs will foster the formation of a fruit containing the seeds that will grow into the next generation. That, of course, is the whole point of flowers.

Before undertaking the challenge of constructing a flower, plants first need to sense when to flower. Many animals, ourselves included, need to mature before they can reproduce, and many then only breed in particular seasons. Plants too need to mature before they can make flowers and produce seeds. Maturity isn't enough, however, to ensure success. Time of flowering matters a great deal, if the fertilization and the subsequent dispersal of fruits and seeds are to be efficient. Many genes are devoted to postponing flowering until the time is right. Among the cues that the time is right are temperature and light, and we will look at both later in the book. Coordinating reproduction with environmental cues isn't restricted to plants. Many animals also respond to cues such as day length. Siberian hamsters, for example, breed during long days, as do some plants.

Unable to move from where they grow, plants have two serious distribution problems to solve if there is to be a new generation. First, they need to have a way to disperse their pollen to either their own carpel or that of neighboring plants of the same species. Later, the plants need a way to disperse their seeds. If all the seeds just drop where the flower forms, newly emerging plants would just crowd one another out as they competed for food and water.

The problem of pollen distribution can be solved by wind, and it's a solution used by some plants. But wind is unreliable and also inefficient because it's not targeted. Evolution found a better solution when it joined the success of plants to the success of insects and birds. The nectar produced by many flowers is an inducement. Insects and birds come looking for nectar and many have evolved shapes that help them to access it, for instance the long beaks of hummingbirds and the proboscis of butterflies. Then, while they are feeding, they pick up pollen as they brush against the stigma and then brush that pollen on the carpel of the same plant or another plant they visit.

Evolution has come up with many different solutions to the second problem, the distribution of seed. The fuzzy balls that contain dandelion seeds are one kind of solution. They are blown by the wind to new places. Gardeners know from the number of dandelion weeds they need to pull up each spring that the process is very efficient. Birds collect other seeds on their feet and feathers and drop them great distances from their source. Animals eat seeds and the fruits that contain seeds and those seeds that are not digested are excreted in new places. This is one of the many natural processes that have made the living world a vast, interdependent network.

Seeds may be very small, but they contain all the DNA necessary to make a new plant, including even a large tree. Half of that DNA comes from each of the seed's parents as delivered by the pollen and the egg. Seeds also contain some food for the new plant to use as it begins to grow, but before its leaves and roots have grown enough to supply sufficient food and water. That's why some seeds, like beans and maize, are large enough to be food for animals, including ourselves.

Chapter 3
SENSING THE ENVIRONMENT

Those who live in mid- to high latitudes, with four distinct seasons, await the appearance of flowers in late winter or early spring, when the days are getting longer and the air temperature begins to warm up. Suddenly, the landscape shows colors instead of drab grey or winter white. Over the next months, the array of blooms, colors, and smells changes in a predictable order, year after year.

Where the change of seasons is not so obvious or even nonexistent, some flowers occur all year long. Even so, the blooming of plants can mark the changing months. Almond trees bloom in late winter or early spring in central California and the Mediterranean region, although any changes in temperature are slight.

The United States Department of Agriculture divides the USA into zones that align roughly with latitude except for the far west. The zones also correlate, more or less, with the minimum winter temperature. Many garden catalogues show the map of the USA marked by these zones. The catalogues' descriptions of plants include recommendations about the zone(s) where each is expected to thrive and bloom. The mid-Atlantic states along the eastern seaboard, for example, are in zone 7, where some varieties of cherry

blossoms usually emerge in early April. Cherry trees bloom somewhat later in zone 6 to the north, which is cooler. In northern
California, zone 9 or 10, they appear even earlier. Wild flowers too
bloom on different schedules.

Anyone in the USA who has been paying attention over the past
few years knows that our temperatures have tended to warm up
earlier than in the past because of global warming. Cherry trees in
zone 7 bloomed in late March in 2012, several weeks earlier than
usual. By the time tourists flooded into Washington DC for the
famous Cherry Blossom Festival in early April, the blossoms were
sad relics of their full pink glory. The Department of Agriculture's
zone definitions were modified in early 2012 to reflect the global
warming trend. The definitions are likely to be modified again in
the coming years if Earth continues to warm. Insects and birds are
more inconvenienced than people by changes in blooming times.
Early flowering can disrupt the essential relation between flowers
and insects and birds if flowers form before their pollinators are
ready to collect nectar.

Temperature is not the whole story. The changing length of
nights (and days) as the Earth makes its annual revolution around
the Sun also influences flowering time. It's not only warmer in
spring than in winter, but the days begin to grow longer.

We all know that days in the northern hemisphere begin to get
longer after the winter solstice in December and shorter again
after the June summer solstice. It's the opposite in the southern
hemisphere where December welcomes the summer solstice and
June the winter solstice. In both north and south, long days are
associated with warmer temperatures especially in temperate
zones. Sunlight in the northern and southern hemispheres comes
at different angles and in different amounts at any particular time

in a calendar year. These differences are the result of Earth's yearly revolution around the sun and its tilted rotational axis. This means that Earth's relation to the sun changes slightly every day. Plants pay more attention to the changes in the timing of the light/dark cycle than we do and those changes also influence the time that flowers appear.

Different species vary greatly in their flowering times. Azaleas and rhododendron show blossoms later in the spring than cherry trees although their buds are visible all winter. Chrysanthemums and asters usually don't bloom before late August or September. This annual progression of color and perfume inspires gardeners to keep on with the hard and sometimes frustrating work of gardening, and a well-designed flower garden will have something in bloom to draw the eye until the first frosts arrive.

This annual, timed display of flowers tells us that plants know a lot about their surroundings. They respond to light, temperature, gravity, and even touch in some cases. And they use this "knowledge" to determine when it is time to begin the process of making a flower.

Annual plants germinate seed, grow, flower, set, and disperse new seed in a relatively short growing season, a matter of weeks. Summer favorites like marigolds, petunias, zinnias, and annual varieties of Arabidopsis flower not long after the seeds are sown in spring and the plant begins to grow. In contrast, some trees grow for decades before they flower. The age and maturity of plants influence flowering time in a way that is characteristic of each kind. Bamboo trees flower rarely and often they die after flowering. Poplars grow for seven to ten years before flowering. Citrus farmers must wait years before their trees are mature enough to flower and produce fruit. The long life cycles make it very challenging to breed trees for desirable features or to use genetic engineering to modify them. Scientists must wait years to learn the results of breeding trees.

It takes seven years for a chestnut tree to mature enough to flower and set seed. That's a big problem for people who are interested in repopulating the United States with these grand trees. American chestnut trees once covered a lot of the country from the eastern seaboard west almost to the Mississippi. The trees were large and beautiful, and collecting their nuts was a favorite pastime for rural youth. They also supported several industries that supplied food for people and animals and lumber for buildings and furniture. Then, early in the twentieth century, the trees began to disappear, victims of a fungal disease called chestnut blight. By mid-century, the chestnuts were gone.

Chinese chestnut trees are resistant to the fungus, so programs were begun to cross-fertilize flowers from the two kinds of trees in the hope that the genes responsible for the Chinese chestnut trees' resistance would become part of the DNA of American chestnut trees. The story is told in a book by Susan Freinkel called *American Chestnut*. Since each generation takes at least seven years, the several generations needed to test the results of cross-fertilizations can take decades. That's a long time to wait. However, with the model plant Arabidopsis, several generations can be tested in a year. Knowledge acquired in the Arabidopsis experiments can be applied to trees. For example, understanding how an annual plant like Arabidopsis changes from the juvenile state to flowering state within weeks has already taught horticulturalists how to hasten the life cycles of trees. With this information, the production of poplar wood and new citrus fruits, for example, may be made more efficient. Perhaps what is learned will hasten the return of the American chestnut tree.

Maturity is clearly an important cue for flowering, but it is not easy for us to observe. Part of plant maturity comes down to the activity of hormones within the plant. A hormone, be it in plants or animals, is a molecule that can travel from cell to cell and influence different cell properties. It acts as a signaling molecule.

Altogether, then, there are at least three different kinds of cues to flowering; one, maturity, depends on internal effects, while the other two, temperature and light, are environmental. Each of the three sets of cues is complex and depends on a chain of events. At the core of these events is the activity of several genes. In addition to the individual complexity of the cues, they all need to be coordinated if the plant is to receive a clear message that it is the right time to flower. We too coordinate environmental and internal signals before initiating many behaviors. Coordination of such signals challenges us. It also challenges plants. Plants sense, process, and coordinate the flowering cues in different ways and on different schedules, depending on their genes.

Understanding the variations in timing of flower formation among different plants depends on understanding how the activities of genes change in response to temperature, light, and maturity. While this is not an easy task for plant scientists, they have had some impressive successes in recent decades. Some aspects of environmental sensing are now fairly well understood, while others are just beginning to yield to experimental work.

Some gene mutations result in plants that flower earlier or later than the typical plant of their species. Some yield plants that fail to flower at all. Studying the effects of a mutated gene can help explain what role the normal gene plays in determining flowering time. One example we have already come across is a mutant plant that makes only leaves and no flowers at all: the mutated gene was dubbed *LEAFY* long before anyone knew what the corresponding healthy gene did, except to foster flower formation. Plants with such mutations are the major tools used to understand flowering. Their seeds carry the same mutations and provide a reliable source of material for experiments. Success in the search for and study of mutant plants requires growing thousands of individual plants. That is where the adoption of the model plant *Arabidopsis thaliana* has been a boon to the study of the genetics of flowering.

Thousands of Arabidopsis plants can be grown in boxes occupying a few square feet of greenhouse space. This facilitates the spotting of mutant plants with unusual flowering habits. Once such mutant plants are identified, their seeds can be collected to provide a reliable source of identical new plants. Special techniques are needed if the mutant plant does not flower at all. Spontaneous mutant plants are quite rare, but the frequency of mutation can be increased if the seeds are treated with certain chemicals or radiation. Also, there are, worldwide, a large variety of wild Arabidopsis plants that show variable behavior; some are annuals, some perennials; some flower early, others late. In recognition of the central importance of Arabidopsis to understanding plants, the entire **genome** of *Arabidopsis thaliana* (the whole sequence of its DNA) was determined in 2000 and is available on the web. Studying the differences in genes between natural variants and mutant plants isolated in the lab is a great help to understanding flowering.

A word of caution. The study of Arabidopsis gives us a framework for understanding how genes influence flowering time and the whole process of flower formation. But different plants have unique flowers and flowering times. We can readily see this from the annual sequence of flowering times in gardens. In temperate zones, cherry trees bloom early as do most crocuses. Then, daffo-dils appear while tulips generally bloom after the daffodils. Each plant demonstrates variations on the themes established by work on Arabidopsis. These variations occur because of differences in the genes in different species; often the changes to the genes are slight and yet result in major changes in the plant's response to environmental cues or the flowers' form.

A lot of what is described in this book reflects study of Arabi-dopsis because many experiments have been done with this exper-imentally convenient plant. Flowering in other plants is often a variation on the themes developed from the study of Arabidopsis. Still, the study of petunias and snapdragons, as well as crop plants like rice and corn, has contributed to what we know…or think we know.

Different plants, like different animals, sense and interpret vari-ous environmental cues in distinct ways. Good food, for example, impinges our senses through a combination of smell, the sizzle we hear from a pan as food cooks, the sight of food on a plate in front of us, the hunger pangs we feel. Each of these cues is perceived by sep-arate systems in our sense organs and brains. But all of them result in our sitting down at table, knives, forks, and spoons lifted, ready for the first bite. Human senses and the ways they are interpreted and coordinated by our bodies and brains are the result of hundreds of millions of years of animal evolution. It's the same with the ways that plants sense, interpret, and coordinate the clues for flowering.

Plants and animals evolve when gene mutations result in new properties that are, by chance, advantageous in an available environment. The offspring of the mutated organism will also carry the mutation and can spread to populate the hospitable environ-ment. New mutations can accumulate and, eventually, a species distinct from the earlier parent, which can no longer interbreed with it, may appear. The current shapes and abilities of existing species are not necessarily the only or the most efficient that we can imagine. Evolution requires only that those shapes and competences work sufficiently well in the environment available to the organism to ensure their survival. If they didn't, different animals and plants would populate Earth. What we see around us are the successes of evolution; the failures disappear. The bottom line for any organism is that genes must work in the surrounding environment if they are to become stable over time.

Plants have evolved the ability to sense whether it's warm or cold, light or dark, and for how long. At this time, research has revealed more about how plants sense light than how they sense temperature. But the way that living things sense and react to light and heat (or dark and cool) is bound by the possibilities that chemistry and physics make available, and both heat and light are forms of energy.

The light from the Sun covers a whole spectrum of wavelengths, of which we see only one portion, and the long-wavelength red end of the spectrum carries less energy than the violet. To sense light and color, plants and animals employ special molecules that recognize and absorb into their structures the energy the light delivers. Different molecules absorb light from different parts of the spectrum, and therefore of particular energies. These complex molecules, like all the molecules in living things, are constructed from simple molecules through the action of catalysts called **enzymes** that are coded for in genes.

Chlorophyll, while it is not a signal for flowering, is a good example of what can happen when sunlight lands on a plant. Chlorophyll is a large, complex molecule that absorbs light from both the red and blue ends of the visible spectrum of sunlight. Green light is reflected, which is why we see plants as green. Other molecules, known as **anthocyanins**, are responsible for the red leaves on Japanese maples. Plants also make use of ultraviolet and infrared light. In each case, some light energy is being absorbed, while the rest is reflected. As we shall see later, the structural changes made in certain molecules when light energy is absorbed can serve as a signal to a plant to do something new, like bend toward the Sun, or make a flower.

While the three flowering cues of light, temperature, and maturity work in different ways, they all result in the activation of a particular plant gene, a gene that starts the flowering process and is called **FLORIGEN**.

More than a century ago, botanists learned that they could force a plant to flower if they injected it with sap from a plant that was producing blooms. The sap contained some sort of "on" signal for flowering. It took another half century or so before the next sentence in the story was written by a simple experiment. A single leaf from a flowering plant, grafted onto the stem of one not yet blooming, is enough to make the recipient begin to form flowers. The two plants could even be different species; plants as distantly related as Arabidopsis and rice both respond to the same signal. These simple experiments gave botanists three important facts: a substance is made in the leaves; it (or a derivative) travels through the sap to the cells at the growing tip where it initiates flowering; and the same substance (or closely related substances) does the same thing in a variety of plants. The mystery substance that causes the dramatic change in the plant's program was named **FLORIGEN**. As a substance that travels through the plant and acts as a signal, it looked likely to be some sort of hormone.

For about 60 years, scientists tried, off and on, to identify FLORIGEN with no success. Then, in the past decade, scientists had the answer: FLORIGEN, the hypothesized hormone that moves from leaves to stems to the growing tip of a plant, turned out to be a protein. Because a protein has to have a corresponding gene coding for it in the organism's DNA, attention turned to the *FLORIGEN* gene and how it gets switched on when it is time for the plant to flower.

Inevitably, the story of flowering is becoming more and more about genes. Before going further with the story of FLORIGEN, we need to delve into what genes are and how they are switched on and off. We will pick up the story of FLORIGEN again in Chapter 6.

PART II

How Genes Work

The rapid development as far as we can judge of all the higher plants within recent geological times is an abominable mystery.

Excerpt of a letter written by Charles Darwin to Joseph Hooker, July 22, 1879

Higher plants might have seemed less mysterious to Charles Darwin if he had known about the work of his contemporary Gregor Mendel. Darwin did not learn about inherited "factors" although they could have strengthened his work on natural selection by clarifying the mechanism of inheritance. We know that Mendel did read Darwin, because there are notes in the monk's handwriting in the copy of *On the Origin of Species* in the monastery library.

About ten thousand years ago, when agriculture began, people knew that if you put a seed in the ground, a plant would grow, flower, and produce more seeds that will grow into the same kind of plant. Implicit in the minds of the earliest planters was the profound idea that seeds contain information, and that seeds from different kinds of plants contain different information. We may think of the "Information Age" as beginning with our invention of computers, but it really began several billion years ago, when life began to evolve on Earth. In the Middle East, early farmers knew that wheat seeds do not grow into rice plants, and in the Americas, that corn seeds do not grow into squash plants. That sounds almost trite, but it's a profound observation. Fast forward ten millennia. After all that time, Mendel opened the door to understanding how seeds can do what they do. It took another century to go from Mendel's factors to establishing the structure of DNA.

Until the middle of the twentieth century, no one knew what a gene actually is. Some biologists thought that genes had to be proteins; others that genes had to be DNA. And the arguments were heated, as they often are when no one really knows the answer. Then, in the early 1950s, biologists showed that a gene is actually a segment of a DNA molecule and that DNA is found in every cell, be it plant or animal. Almost simultaneously, the beautiful, double-helical molecular structure of DNA molecules was described by Francis Crick and James Watson, and it quickly became a twentieth-century icon.

These discoveries were made when I was a beginning graduate student studying biochemistry. They totally changed biology for me and for all biologists. They began an exciting period of research that has gone on for more than sixty years and shows no sign of flagging. Each fresh discovery reveals unsuspected new facts and questions about living things.

Chapter 4 describes how DNA molecules embody the information in genes. The way genes function follows quite logically from the structure of the DNA molecule itself. Plants and animals can have about 20,000 genes in their genomes, but only a selection of those genes may be needed at any given time or place (for example in roots, stems, leaves, or flowers) in the organism's life. Mechanisms for turning genes on and off are essential for biological success and many genes function as on/off switches for other genes (Chapter 5).

The chapters of Part II may, at times, seem more technical than some readers anticipated. And they are not particularly related to flowers. But these chapters lay the groundwork for understanding the when and how of flower formation.

Chapter 4

HOW GENES WORK

If you ask Google for a definition of the phrase "information system," the computer returns a long list of sources. Everything on the list relates, one way or another, to how humans or computers obtain, store, and transmit information. Fortunately, James Gleick, in his landmark 2011 book entitled *The Information*, takes a more comprehensive view. Gleick's definition includes scripts, African drum language, and DNA. Not only does DNA store genetic information in the form of genes, but that information is transmitted with remarkable fidelity from one organism to its progeny, and from parent cell to daughter cells at every cell division.

The information in DNA includes on/off switches that control when and where in the organism's life individual genes are used. For example, genes required for the construction of flowers are turned off when the young plant is growing. Genes needed to make seeds are only turned on when the flower is formed. The system works the same way in all living things, and while some information is common to all, some is limited to one or another kind of organism. Animals, for example, don't have the genes that enable the making of chlorophyll, the green plant pigment. Many of the differences between plant types (and similarly between various animals) depend not just on what information is there, but on how and when the information in DNA is used.

What is DNA? To begin, it's a molecule, which means it is made up of atoms as are all things in the universe. Each DNA molecule is a long chain made of repeating units of only four similar arrangements of atoms we'll call **bases**. It's like a long chain of beads of four different colors. We needn't bother with the chemical structures of the bases or even their proper names; they are often just called by their initials, A, G, C, and T. The long polymer chains of linked As, Gs, Cs, and Ts are held together by identical chemical links between neighboring bases. The information content of DNA lies in the order of the four bases along the chains, just as the information in words is in the particular letters and how they are arranged. Thus, the sequence ATGGTA encodes different information from GTGTAA.

I have already touched on the difficulties of defining a gene. But for most purposes, a gene may be simply defined as a heritable unit of biological information. Genes are usually segments of DNA of a few tens, hundreds, or thousands of bases.

The three-dimensional structure of DNA is crucial to its function. It is a "double-helix," consisting of two strands wound around each other, with the bases interacting with each other as pairs: A only pairs with T, and C with G. Thus, if a segment of DNA has the base sequence GAAGATCT, the other strand in that segment will be CTTCTAGA. The two strands are said to be complementary. Genes may occur on either of the two DNA strands.

DNA is so important that, like a super-guarded head of state, it is protected from exposure to much of what goes on in plant and animal cells by being restricted to the cell's nucleus. By itself, DNA doesn't do anything in cells except carry instructions, including those to build the proteins and other molecules that do the actual work of the cell. Copies of the base sequence of individual genes are used to give the gene's instruction to the cell. The copies are produced in the form of a related molecule called RNA, which can go wherever in the cell it is needed.

RNA chains are much shorter than DNA because they are copied off only segments of DNA; usually the segments correspond to the length of a particular gene. While DNA molecules often have millions of units, RNA molecules have thousands and fewer; some are as short as a few bases long. The bases in RNA chains are linked to a different kind of sugar than that in DNA, and the base T in DNA is substituted by a similar base, U, in RNA. The bases in RNA are then A, G, C, and U.

Most of the time, RNA is copied off only one strand of a segment of two-stranded DNA and is therefore a single chain. Copying into RNA obeys the base pair rules, so that an A on the DNA is a U on the RNA; a T on the DNA an A on the RNA; a G on the DNA a C on the RNA; and a C on DNA a G on RNA. If the segment of DNA being copied has the base sequence GAAGATCT, the corresponding RNA sequence will be CUUCUAGA. The copying is accomplished by protein enzymes that, like all proteins, are themselves encoded by genes.

There is one big exception to the isolation of DNA in the nucleus. Every time a cell divides, as they do in a growing plant or a developing flower, the DNA must be copied so that each daughter cell receives a faithful copy of the DNA and thus all the genetic information. A dividing cell must open its nucleus so the two copies of the DNA can be divided between the two daughter cells. Each of the many cells in plants (and animals) has a copy of all the DNA that was in the seed and thus all the genes common to that organism.

The copying process is, like copying into RNA, largely accurate. DNA copying too is accomplished by enzymes that are themselves encoded by a gene.

Seeds represent a new generation, and in them and in all plant cells, half the DNA comes from the maternal parent (in the egg) and half from the paternal parent (in the pollen grain). This is the same thing that occurs when animal egg cells are fertilized by sperm. Offspring contain a combination of information from each parent.

Plants and animals devote a lot of their resources to keeping DNA safe and unchanged. Most of the time they succeed, and new cells and even whole new offspring receive true copies of parental DNA. But sometimes, when DNA is copied before cell division, mistakes are made. If copying goes badly awry, a new cell or even the whole plant may die. Sometimes, a change is minor and the DNA passed to a new cell is only slightly different from that of its parent cell. If the changed segment is associated with a gene, the properties of the cell can change in major or minor ways. Mutations in the pollen or egg DNA will be passed on to new generations and the offspring will be different from its parents.

Mutations can spell either trouble or opportunity. Cystic fibrosis and hemophilia are human diseases that are the result of mutations in DNA. Mutations also appear to be the reason that a few people are naturally resistant to HIV. There are plant mutations that result in stunted growth or flowers that form earlier or later than they should. Other mutations have given rise to flowers whose sepals resemble petals, as in orchids and crocuses. But mutations were also important to our astute ancestors, as they bred wild animals or wild plants looking for improved varieties. Mutations in the wild tomatoes growing originally in the Peruvian Andes produced the many varieties of the now popular fruit. The selection of mutant varieties of a wild Mexican grass gave us corn with its sweet kernels. Plant mutations give us ever expanding colors in our gardens. Many flowers that people admire and seek out for their gardens are the result of breeding plants that had chance mutations. Wild roses, for example, have only a few petals, not the lush abundance of petals we see in cultivars in our flower shops and gardens.

DNA can change in a variety of ways to generate mutant genes. Some mutations are a change in the sequence of bases in a gene coding for a protein. In others, a few bases or even the whole gene sequence might be lost from a DNA segment. Or DNA segments may be repeated rather than omitted. Repetition can be just as deleterious to the organism as a deletion. But repetition can also provide the raw material for a new gene if the repeated DNA segment itself becomes mutated and evolves to fulfill a new function. Still other mutations occur when a moveable DNA segment, what is called a jumping gene, inserts itself into the middle of a gene. Mutations in genes whose products switch other genes on or off have been especially helpful in understanding what happens during the development of plants and animals.

There are two kinds of gene. Both work by first copying the sequence of DNA bases into RNA segments. In one kind of gene, **RNA genes**, RNA is the final product; the RNA molecules may help to switch other genes on or off, or they may modulate the activity of other genes, or participate as enzymes or scaffolds for cellular events. The second kind of gene is copied into what is called **messenger RNA** that then instructs the cell to manufacture particular proteins; these are **protein genes**.

There is, though, rather more to DNA than strings of bases neatly making up genes that code for proteins and RNA. Some genes are made up of several segments of DNA separated by non-coding segments of DNA called **introns**, which need to be removed from the RNA copy before the sequence can be functional. Other segments of DNA contain important information for controlling when a gene is active or inactive and may not be copied into RNA. Still other important segments don't code for anything but define and protect the ends of long DNA molecules. Finally, there is a huge amount of DNA, perhaps as much as 90 to 95 percent of the total, whose function is presently unknown. Just what all that DNA does, if anything, remains to be seen.

Proteins are folded molecules made up of varying length strings of twenty different chemical units called **amino acids**. Each protein has a particular number of amino acids arranged in a characteristic order; their lengths vary from tens to thousands of amino acids. The DNA in each protein gene provides instructions for the formation of a particular protein with a specific sequence of amino acids. The big challenge for biology in the early 1960s was to figure out how the four bases of DNA could determine the ordering of the twenty amino acids that specified the protein.

Some scientists approached the challenge as a genetic problem. They learned that three DNA bases defined one amino acid, but nothing about which bases matched which amino acid. While this work drew a lot of attention because it was done by well-known and respected scientists, a young and then unknown biologist called Marshall Nirenberg took a biochemical approach. He surprised everyone by proving that the RNA triplet UUU in the messenger RNA is the signal to insert the amino acid called phenylalanine into a protein. This meant that the DNA gene had the sequence TTT at the equivalent place. This was the first triplet, or **codon**, determined in what is called the **genetic code**.

The surprise was announced in a dramatic way. The 1961 International Congress of Biochemistry was held in Moscow. At the time, Mendelian genetics was unpopular in the USSR for political reasons, and some geneticists were in jail. Nirenberg, then thirty-four years old, was allotted ten minutes for his talk. It was the amount of time usually given to ordinary work. The few people who heard him began to talk about his exceptional announcement. The next day he was asked to repeat the talk in a large hall filled to capacity. The discovery was instantly recognized as a huge advance in biology. Nirenberg and his team went on to determine the codons for the other twenty amino acids in proteins.

Here is how the genetic code works. Each sequence of three bases in messenger RNA directs the cell to put a particular one of the twenty different amino acids into the protein that is being made. For example, the sequence UUUGUC instructs the insertion of the amino acid phenylalanine (UUU) followed by the amino acid valine (GUC) into a growing protein chain for the segment phenylalanine-valine. Those DNA regions constituting a protein gene contain a series of such three-base codons (with T instead of U).

You may have noticed that there are sixty-four ways to arrange the four bases in groups of three, but only twenty amino acids, and may be wondering about what happens with the other forty-four codons. It turned out that many amino acids match more than one codon. Phenylalanine, for example, matches UUC as well as UUU. Valine matches four codons: GUU, GUC, GUA, and GUG. The long DNA chains (and their RNA copies) contain only the four bases. There are no separate chemical punctuation marks indicating the beginning or end of a gene's protein sequence. Instead, certain codons serve as punctuation. The codon ATG matches the amino acid methionine but also, when in a special context of bases, marks the beginning of a protein coding segment of DNA (AUG in RNA). Three codons don't match any amino acid but mark the end of the protein coding segment: TAG (UAG), TGA (UGA), and TAA (UAA). Since proteins come in different lengths from tens to thousands of amino acids, the genes that code for them vary in length accordingly.

A change in the DNA of a protein-coding gene results in a corresponding change in the messenger RNA that then changes the amino acids in the protein. If, for example, the segment TTTGCT were to change to TTTTAT, the RNA copy would be AAAAUA rather than AAACGA and the amino acid sequence would be lysine-isoleucine instead of lysine-arginine. This may have no effect on the way the protein functions, or it may make the protein inactive, or it may alter the way the protein does its job. But if TTTGCT was changed to TTTTAA, the RNA copy will be UUUUAA and the protein would not be extended past the UUU (phenylalanine) because of the stop codon, UAA, and likely not function at all. This is the kind of change that can occur during the copying of DNA before the cell divides.

Once the genetic code was identified in the early 1960s, it was, in principle, possible to take DNA sequences and predict both the RNA sequences of RNA genes and the amino acid sequences encoded in protein genes. In principle, but not in practice. The problem was that there was no good way to determine the sequence of bases in a DNA segment or even to isolate pure segments of DNA. It took more than a decade before the sequencing of DNA segments became a task that any laboratory could undertake. Two technological advances marked this big advance. One was the ability to isolate and purify particular DNA segments out of the long chains in any organism, the technique known as **molecular cloning**. The other was the invention of fast and efficient procedures to determine directly the sequence of bases in a purified DNA segment. In my own lab in the mid-1970s, a colleague and I spent the better part of eighteen months in tedious work determining the sequence of a 172-base-long DNA segment. Just as we finished, a paper describing the new sequencing technique was published; it took about six weeks to set up the new method and confirm the sequence of the segment. And sequencing has become much, much faster since then.

Many proteins are enzymes that act as catalysts to accelerate the chemical reactions required to support the functions of cells. Other proteins bind to DNA to regulate the switching on and off of genes. Still other proteins build the structures of the organism, like roots, stems, and leaves. And one group of proteins, the **histones**, is intimately associated with DNA and the activity of genes, as we shall see in the next chapter. An organism's DNA, its genome, contains genes with the instructions to build all these proteins.

The many RNA genes in DNA are copied into RNA molecules, each of which contains a characteristic number of bases. Such RNA molecules are important to cells without ever coding for a protein. These RNAs help perform many of the essential functions of a cell, including splicing out the introns to make a useful messenger RNA. Some RNAs are needed to construct and operate the cellular machines that translate the codons in the messenger RNA into proteins. Many other RNA chains are switches that influence what, how, and when other genes are to be used in the life of the plant (or animal). Some of these play critical roles in the flowering process, and they will be an important part of the stories in the rest of this book.

Today, the sequences of billions of bases in DNA segments from many organisms including many plant genes are known and stored in huge data banks; these banks are readily accessible to all through the Internet. The DNA data banks include, for example, the whole genome of that rare flowering plant *Amborella*, thought to be related to the earliest flowering plants that grew on Earth. Convenient and powerful search programs allow anyone to search a genome for a sequence of interest. With a protein gene, a short computer command translates the sequence of bases into the sequence of amino acids in the corresponding protein and provides information about the function of the protein and much else.

Cells in Arabidopsis each have about 270 million bases (As, Gs, Cs, and Ts), divided between five pairs of **chromosomes**, which are the organism's characteristic packages of DNA and protein (or 135 million bases in each set). For comparison, humans have twenty-three pairs of chromosomes, and a total of 3 billion base pairs in each set. The full set of chromosomes occurs in every cell and is something like a multivolume encyclopedia, each book of which contains a different collection of facts in the form of genes. In all sexually reproducing organisms, one of each pair of chromosomes comes from the female parent and the other from the male parent. The pairs carry redundant information except for relatively small differences between them. It's as if you owned two different editions of a multivolume encyclopedia.

Regardless of the large difference in the total number of bases or chromosomes characteristic of each organism's genome, estimates of the number of genes in Arabidopsis and humans are very similar, between 23,000 and 25,000. The larger size of the human genome, about twenty-five times that of Arabidopsis, is caused mainly by a very large number (ranging from a hundred to thousands of base pairs in length) that are of uncertain function, if any, and are repeated many times in the genome. Some of those repeats occur in long tandem arrays, while others are sprinkled throughout the genome.

This chapter is the beginning of a two-chapter-long diversion from between the end of Chapter 3 and the beginning of Chapter 6 where the story of FLORIGEN is picked up again. For decades FLORIGEN was known only as a mysterious substance that can turn on flowering. Sixty years of effort were unsuccessful in identifying the nature of FLORIGEN. Now, we know that FLORIGEN is a protein that is made in leaves and travels up stems to the growing tip of a flowering plant. Once at the tip, FLORIGEN starts a cascade of events that result in the development of a flower. Because FLORIGEN, like all proteins, is coded for by a gene, attention turned to what turns the *FLORIGEN* gene on when it's time to flower.

Turning genes on and off at the proper time and in the proper place in plants and in animals is the key to orderly development of whole organisms and their individual organs, including flowers. It's important enough for flower formation that it needs a chapter to itself.

Chapter 5

SWITCHING GENES
ON AND OFF

D NA and the genes it contains give plants much of the information they need to grow, flower, and make seeds and fruits. All
these events need to happen in the correct order and at the appropriate place and time. After all, a plant cannot make a fruit before
it has a flower. Nor does it grow leaves on its roots. This means that
the plant's genes are used selectively with respect to both time and
place. A set of closed encyclopedia volumes sitting on a shelf or in
the Cloud is not very useful. The volumes would be equally useless
if all of them and all their pages were open all the time. An encyclopedia is only useful when a reader chooses to look up a particular
subject. It's the same with DNA and genes. Some genes are needed
when the seed begins to sprout. Others go to work as the seedling
emerges into the light; these include the genes that direct the formation of the green pigment chlorophyll. Still other genes, like
FLORIGEN, are used later, when it's time to make a flower. And yet
another set of genes is activated to make a fruit containing seeds
making up the next generation.

Deciding when to open an encyclopedia volume and which volume to open is a complex task for our brains. How does a plant
"know" when to turn off one gene and turn on another? And how

are the switches operated? What do we know about the switches that turn plant genes on and off in the proper organ and at the correct time? To start with, we know that the switches consist of either RNAs or proteins that are encoded in the plant's own genome. Many of the genes described in this book are genes for switches. What switches *those* genes on? Probably other RNA and protein switches. How does it all get started?

Seeds themselves are dried out. With no water, all the usual cell processes stop. But the RNAs and proteins that were formed as the seed was maturing are still there. When the seeds become wet, they are ready to be active again. Those who live near deserts see this for themselves. Soon after the first big spring rain, the desert comes alive with plants and flowers. The same thing happens when the dry seeds we purchase in shops are planted and watered. In a few days, stored proteins and RNAs are activated and begin to do their work. New cells form, and a root and shoot soon emerge. From then on, the plant and its DNA are on their own, as long as some water and a few nutrients are available. The RNA and protein the plant obtained from its parents are the switches that renew life in the seed.

Different kinds of cells such as those that form stem, root, leaf, and flower switch on different sets of genes. In addition, all the cells switch on a common set of "nuts and bolts" genes. Included in the "nuts and bolts" category are the genes for special proteins and RNAs that constitute the machinery for making more DNA, RNA, and proteins. Other genes are switched on (and off) at different times in a cell's life. It's something like a cabinet of pots in a kitchen; not all of them are used to make every meal, and some are never used even though we keep them, just in case they come in handy in future. This means that each cell can act quite independently and do many things on its own. It can also do many things in coordination with other cells, if it receives appropriate chemical signals.

Before we can think about how the switches actually work, it helps to have some idea of the way DNA occurs inside a cell nucleus. Recall that the sequence of bases on one strand of the DNA double helix is matched, on the other strand, by the complementary sequence: As and Ts are matched, as are Gs by Cs. In each different segment of DNA only one of the two strands usually makes genetic "sense," that is, has a sequence that can be translated into a protein using the genetic code or to construct the proper RNA. Most often, the other strand is, in that region of the long DNA molecule, nonsense. Switches that turn genes on or off also need to recognize the difference between the strands. For a gene segment, one strand may attract a protein or RNA switch while the other does not. The switches need to ensure that the correct strand gets copied into the correct RNA or the whole system would have fallen apart before it even got started.

There are a lot of switch proteins and RNAs that are important for making a flower. Most often they operate by interacting with DNA base sequences just ahead of the start of a gene sequence. One typical DNA recognition sequence for a plant switch molecule is CACGTG; the plant hormone abscisic acid interacts with this base sequence and influences the activity of the neighboring gene. The presence or absence of a switch molecule near a particular gene dictates whether that gene is on or off. In many cases, there are several such recognition sites that stimulate or inhibit the activity of a single gene. This means that multiple proteins or RNAs or other molecular switches can influence whether the neighboring gene is on or off, and whether it is on full blast or only at a low level. Of course, there are switches that determine when and where genes that code for other switches are on or off. Not only that, but some switch molecules can turn on one gene and turn off another!

The switch molecules—proteins, RNAs, and hormones—can control where and when particular genes are turned on or off. But for any of these molecules to act as a switch, it needs access to the DNA in the region of the gene it is controlling. It's like needing to be standing at your front door before your key does you any good.

Why is access to DNA an issue? That's a question that has only been asked and answered in recent years. It turns out that two long-known but previously puzzling features of DNA are involved in blocking or providing access. The first puzzling feature was the presence on some, but not all, of the C bases in DNA of a small chemical appendage called a **methyl group**. (A methyl group, CH_3, is one carbon atom attached to three hydrogen atoms and is a close relative of the gas methane, which has four rather than three hydrogen atoms attached to one carbon atom.) The methyl groups do not seem to interfere with the working of the genetic code or with the pairing of a C with a G in a double helix. Yet most chemicals in cells are there for a reason. What were the methyl groups doing?

It turned out that marks like the methyl group on C bases and a few other small molecule appendages that occasionally occur on DNA influence the activity of genes without changing the base sequence. They accomplish this by barring access to the gene by switch molecules and/or by the machinery needed to copy the DNA into RNA. They make the gene inaccessible. Genes that have such appendages on their own or neighboring DNA bases are silent.

The methyl and other marks on DNA bases are called **epigenetic** marks: the "epi" prefix stresses that they provide information in addition to typical genetic information encoded in the sequence of

DNA bases. Epigenetic marks are added to DNA bases by special enzymes called **epigenetic writers**. DNA in different kinds of cells can have different epigenetic marks. Epigenetically marked genes are recognized by proteins called **epigenetic readers** that identify the modified base and stop the activation of the marked gene. When it is the right time and organ for the gene to be activated, the mark(s) are removed by other special enzymes called **epigenetic erasers**. (These names—epigenetic writers, readers, and erasers—are convenient, but don't imply anything about how these enzymes work.)

A cell that acquires (or loses) an epigenetic mark can pass the mark (or its absence) to its offspring cells, which means it's part of the genetic heritage of parent and offspring cells; that's why the term "genetics" in epigenetics is apt.

The second surprise about access to DNA stemmed from better understanding of the role of the proteins called histones. Histones are intimately associated with DNA, and they have two kinds of job; one concerns the structure of DNA in cells, and the other the function of genes. The structural job is to help DNA to fit inside a cell. The functional role is to participate in the switch processes that activate and silence genes.

DNA strands are much thinner than a hair; they measure about a fifty-millionth of a meter in diameter. In contrast, they can be as long as a meter in length. Cells typically measure about a tenth to a hundred thousandth of a meter. It wouldn't do to have all that information in a big, spaghetti-like mess inside cells. Evolution solved that challenge by neatly winding DNA chains around protein scaffolds. The winding occurs in several stages that result ultimately in the compact DNA–protein packages known as chromosomes, that fit easily inside cells.

In the first stage of the winding, DNA chains loop around clusters of histones; a loop takes up about 150 units of As, Ts, Gs, and Cs. The clusters of eight histones associated with each DNA loop contain two molecules of each of four distinctive histone proteins, and the whole loop of DNA around the cluster of eight histones is known as a **nucleosome**. Between the loops are short, "spacer" DNA segments of variable length that can be associated with a fifth kind of histone. The result, when seen in a powerful microscope, looks something like beads on a string. The combined packaged structure of proteins and DNA is called **chromatin**. Histones were recognized more than 100 years ago and very similar histone molecules occur in both plants and animals. Until quite recently biologists held two ideas about histones that proved to be wrong.

One misconception was to recognize only five different histone proteins. There are additional histone proteins that associate with chromosomes in special places and times and help determine whether a gene will be active or inactive in a particular kind of cell or at a particular time in the life of the plant. The variant histone molecules have slightly changed amino acids from the canonical ones and are encoded by distinctive histone genes. These different histones can play an important role in determining whether a gene is active or inactive in a particular kind of cell or at a particular time in the life of the plant. One of these histones winds the DNA especially tightly on the "beads," tightly enough that no other proteins can get near enough to turn on particular genes.

The second misconception was to assume that histones are just inert scaffolds for winding up the DNA. In fact, they can control access to the start of many genes, thereby either allowing or barring association with proteins needed to turn the gene on or off.

Dynamic modifications of the amino acids on histone chains are a major factor in regulating the activity of nearby genes.

While the role of histones in regulating gene activity was discovered relatively late in the unfolding of the genetic story, histone modifications are of central importance in both plants and animals. How do researchers learn about such subtle mechanisms? As so often the case, the understanding has come from analyzing mutations. Normal histone modifications are made by enzymes and problems ensue when mutations occur in the genes that encode those enzymes. Cancer cells, for example, frequently carry mutations in the genes needed for appropriate modification of histones. Plants carrying mutations that effect histone modifications are the basis for understanding many aspects of flower formation.

Some dynamic modifications of histone proteins involve adding or removing a methyl group on particular amino acids on histone molecules. Depending upon which amino acid on which histone bears one or more methyl groups, the neighboring gene may be silenced or activated. One example that plays a role in determining when a plant will flower concerns the histone called histone 3. The amino acid lysine occurs multiple times in the histone 3 molecule. When the lysine at position 4 of the protein's amino acid chain is decorated with methyl groups, the gene in that region of the DNA next to it is accessible and tends to be turned on. But if instead, methyl groups decorate the lysine at position 27, the gene will not be turned on by its switch protein and/or RNA. These differential effects of methyl groups on histone 3, depending on whether they occur on lysine 4 or lysine 27, hold true for many genes.

Other small chemical additions to the amino acids in histones have their own effects on neighboring gene activity; these additions include chemical groups like phosphate or acetate. In some cases the effects of acetate groups are the opposite to those of methyl groups. None of these changes alter the basic sequence of amino acids in histones that are determined by the histone genes. They can, however result in subtle changes in the shape of a folded histone molecule around which DNA is wound. A concomitant change in the winding of the DNA can make it more or less accessible to the switch proteins needed to turn a gene on or off. Now, biologists are actively studying the role of histone modifications in influencing gene action. Like the genetic code itself, the effects of histone modifications are similar in plants and animals. This indicates that the histones and their modifications have roots in early evolution, before the separation of the branches of animals and plants.

The more we learn about how genes are turned on and off in a cell, the more complicated the story seems to become. The information inherent in the extra chemical groups on DNA and histones really matters. The role of some of these marks will become clearer later in this book when we consider examples of how the histones and their modifications affect the formation of flowers. If this book had been written a decade ago, histones would probably not have been mentioned except as inert, purely structural scaffolds for the DNA in chromosomes. Now, we know that they are active players in determining what goes on in plant and animal cells. Like the epigenetic changes that are made directly on DNA (see the earlier discussion on epigenetic marks), the modifications of some histones also seem to be inherited by daughter cells upon cell division. In fact, some people also use the term epigenetics to include the histone marks that influence gene activation. Other scientists hold that the term epigenetics should be reserved for the marks on DNA. Like the discussions about the definition of a gene, this discussion doesn't affect what we actually know or understand.

Perhaps biologists should have recognized sooner that hidden away behind some unexplained observations, there was an information system beyond the genetic code. That oversight held up scientific understanding but also had nonscientific consequences.

More than a decade ago, I joined the board of what seemed an exciting new start-up company. Several investors believed the same thing and put millions of dollars into the company. I too made an investment, albeit a modest one, because I believed in the potential of the genetics the company would exploit. But the company failed and we all lost our money. The failure was in part because we did not yet appreciate the importance of the small chemical marks on DNA and histones. Similar failures doomed other start-up biotech companies. It's best always to recognize that our knowledge may be incomplete.

We now know that the chemical marks on both DNA and the histones around which DNA is wound control, in at least some instances, the access of switch proteins to genes. Usually the switch proteins that turn genes on and off associate with the region of the gene just ahead of the position where a coding region begins. One thing the switch proteins do is give access to the enzyme needed to copy the DNA into RNA. It's like having multiple doors to a house. One key, changed marks on histones, opens the main front door; a second key, a switch protein, opens an inner door so that the enzyme that copies the DNA into RNA can gain access to the gene. The DNA genetic system is specialized for establishing, passing on, changing, storing, and utilizing information, while systems such as histone modification and epigenetic marking of DNA regulate access to and modulation of that information.

Given a congenial environment, this messy process involving complex, interacting networks that regulate genes works to great effect, directing organisms to grow and reproduce with some efficiency. This is just what we might expect from an evolutionary process that is undirected, and involves random mutation followed by survival only of those organisms to which the changes confer an advantage, or at least do not prove detrimental to survival and reproduction. The regular appearance of flowers is testament to the success of gene regulation, and the power of natural selection in assembling such a process, over vast numbers of generations, through the weeding out of the less successful among random mutations.

The central point of this chapter is that individual cells or related groups of cells do not use every available gene at all times in the life of a plant. Complex regulatory networks dictate which genes will be active in each cell and tissue, and when. Different genes are switched on in the roots and shoots of plants. And when it is too early for a young, growing plant to flower, the genes needed to make flowers are kept switched off. How a cell behaves depends on a cascade of signals from within the plant or from the environment.

PART III

Time to Flower

The Flowers that bloom in the spring, tra la
Bring Promise of Merry Sunshine
As we merrily dance and we sing, tra la
We welcome the hope that they bring, tra la
Of a summer of roses and wine.

W. S. Gilbert, *The Mikado*

In Chapter 3 we noted that the coordination of the flowering cues—light, temperature, and maturity—instruct the plant to initiate flower formation. This instruction leads to the activation of the gene for the protein FLORIGEN in the leaves. Chapters 4 and 5 gave some background on the nature of genes and how they are switched on and off at appropriate times and places in a plant (or animal). Now, it's time to return to the process of flower formation and in particular to how the *FLORIGEN* gene is switched on.

Chapter 6 describes the role of maturity and hormones in flowering. Maturity is necessary but not always enough to ensure flowering. In some plants, maturity means that they are ready to sense and act on environmental cues like length of day and temperature. In other plants, maturity is sufficient to induce flowering independent of environmental cues; signals within the plant initiate flower formation. The subsequent chapters describe what plant scientists have learned about the way plants respond to environmental signals that are important for flowering. By the end of Chapter 8, everything will be ready to begin the formation of a flower.

Chapter 6

GROWING UP GREEN

It's time to return to the protein FLORIGEN and its gene, which were introduced at the end of Chapter 3. FLORIGEN protein is made in leaves and then travels up the plant stem to the growing tip of the plant, the meristem. When it arrives at the meristem, FLORIGEN is the signal to initiate flowering.

Environmental cues are not likely to promote flowering if the plant is too immature to reproduce. Gene products and various processes within the plant cells themselves work to ensure maturity. Several of these processes and genes were known long before anyone appreciated their significance. Consequently, they were grouped together and called "autonomous factors," because they promoted flowering independent of environmental conditions such as light and heat. In a way, the name was an acknowledgement of ignorance.

Now we know that some of these "autonomous" genes are involved in the maturation of the plant. Others are responsible for making plant hormones or are RNA genes that can affect both maturation and flowering. Still others influence epigenetic processes, including histone modifications and chromatin structure. Perhaps they are better called "internal" rather than "autonomous" signals, to stress their difference from environmental signals such as light and temperature.

Often, it's possible to tell whether an annual plant is mature enough to reproduce by counting its leaves. The size, the number, and, sometimes, changes in the shape and texture of leaves can all be indicators. Biologists generally use the number of leaves on a young annual plant as a reasonably reliable measure of maturity. Presumably, a plant needs a certain number of leaves to provide the energy and materials required to make a flower.

That's not to say that there is some magic number of leaves for all plants. Arabidopsis is ready to flower with a dozen leaves or so but the five-foot high lily plants in my late spring garden can have twenty-five leaves or more and still not be ready to yield their brilliant yellow, heavily scented blooms. Trees, of course, have innumerable leaves before flowering; poplars, as we have seen, take seven to ten years to mature.

The time from the sprouting of a seed to maturity can vary even among different varieties of the same annual plant. Some gene mutations in Arabidopsis and maize can speed up or slow down the maturation process. This simple observation tells us that when such genes are in their normal state, they and their protein or RNA products function either to help maintain the juvenile state or to signal that it's time to mature. The discovery of these genes is a good example of how useful Arabidopsis is for probing the systems that govern flowering time and flower formation.

Experiments are just beginning to reveal hints about the processes plants go through to mature. A sugar called trehalose begins to collect in leaves and meristem as a new plant ages. Perhaps the accumulation of such nutrients contributes to a mature condition. Other hints suggest that the appearance of small RNA molecules, coded for by RNA genes, is another sign of maturity in perennial plants; such RNAs could control aspects of gene activity associated with maturity.

For example, there are changes in the activity of two particular RNA genes that accompany maturity in annual Arabidopsis and its wild relatives. The RNA produced from one RNA gene seems to be necessary and sufficient to keep a plant in its immature condition and inhibit flowering. The amount of this RNA in the plant goes down as the plant grows and forms more leaves. The amount of the second RNA gene product follows the opposite trend; its amount increases as the plant grows, forms more leaves, and prepares to flower. The net result is that after about five weeks of growth, the plants are able to respond to environmental conditions and become capable of flowering. More research is needed to explain just how these RNAs work.

Certain plant hormones also influence plant maturity and ability to flower. Some of these chemical messengers are proteins, while others are small molecules that are neither proteins nor RNA. Hormones can be distributed throughout a plant or animal rather than being restricted to one kind of cell or tissue. Estrogen, testosterone, and insulin are examples of animal hormones. Two plant hormones important for flowering are **auxin** and **gibberellin**. Both of these hormones have multiple effects on plant growth, architecture, and development as well as flowering.

The structure of auxin is related to that of the amino acid tryptophan, from which it can be made in plants. The gibberellin molecule is quite different, and is related to molecules called terpenes that give some plants their distinctive scent. Both hormones can promote the activity of *LEAFY* and several other genes in meristem cells even in the absence of environmental cues. It is possible that gibberellin also acts indirectly on flowering time, because it is able to stimulate the amount of a protein called PIF4, which in turn binds near the FLORIGEN gene and activates it, as we shall see in the next chapter.

If gibberellin stimulates the manufacture of enough LEAFY protein, flowering can occur even in the absence of FLORIGEN. This is most easily seen in plants that flower when days are short, such as chrysanthemums. In plants that flower when days are long, there is more time for other contributors to flowering such as temperature and light to work, and the effect of gibberellin is not so obvious.

LEAFY is not the only gene switched on in the meristem by gibberellin. The hormone also switches on genes that would be activated in response to the arrival of FLORIGEN. But if both the FLORIGEN and LEAFY genes are absent or inactive, the plant keeps on making leaves and no flowers appear. Such mutant plants are helpful for experiments but are not likely to survive in nature since without flowering they cannot produce seeds.

Gibberellin doesn't need to be made within the plant to be effective. Plants that are immature or have a mutation that stops gibberellin synthesis can flower if the hormone is added externally. This is one of the tricks used by nurseries and gardeners to induce flowers in young plants. It's easy for gardeners to buy gibberellin; it's advertised and sold for these purposes. A solution of a gram of gibberellin in ten gallons of water is enough for a huge garden.

There are other genes too that influence flowering in ways independent of environmental factors. Annual plants go from seed to plant to flower to seed in the space of a few months. Typically, they have low amounts of FLOWERING LOCUS C or FLOWERING C, a protein that stops early flowering. But some annual plants accumulate a lot of FLOWERING C because of additional mutations that regulate FLOWERING C gene activity on their own.

Some of these mutant plants turn the activity of the *FLORIGEN* gene down or off. The mutant genes, some of which are RNA genes, might do this in ways similar to the RNA genes involved in reducing *FLOWERING C* activity after periods of cold, something we will look at in the next chapter. Other mutations involve the epigenetic modification of the histones in the chromatin associated with the *FLOWERING C* gene. Altogether, the normal action of these genes is to turn on the *FLORIGEN* gene and promote the transport of FLORIGEN to the meristem.

It may seem that the story that was just told is a complete, neat explanation of how a plant becomes capable of flowering. However, there is still a lot to be learned about flowering and new information can change the story in small or large ways. It's important to remember, too, that most of the relevant experimental work has been done with Arabidopsis. It is certain that other plants, while depending on similar genes and processes, may also use them in unique ways.

The stories of LEAFY and FLORIGEN and their genes have just begun. Once plants are mature enough to be capable of flowering, environmental factors such as temperature and light work together, along with a large number of additional genes, to ensure that flowers actually form.

Chapter 7

WARM AND COLD

Once a plant is mature enough to flower, environmental influences take control of the process. This chapter is about the effects of temperature on the activity of the *FLORIGEN* gene, and we'll come to the influence of light in the next.

While it is obvious from the reactions of plants to warming weather that they sense the surrounding temperature, science has not yet come up with a clear understanding of how a plant knows its surroundings are warming up or cooling down. Even an increase of two degrees in temperature over normal can stimulate a plant or tree to flower days before it is expected to. That's why so many flowering trees and bushes have bloomed early in the recent warm springs. Small increases in atmospheric temperature turn up the expression of the *FLORIGEN* gene and hasten flowering. This kind of response to warming (and the opposite response to cooling) appears to occur in both annual and perennial plants. As our planet continues to warm up, we can expect other changes from what used to be the usual schedule for flowering.

We also know very little about how the messages a plant receives about the surrounding temperature are translated into the responses we see. A few observations hint at the possibility that plants sense temperature in a way related to that by which they sense light. This is not such a far-fetched idea, since heat energy is infrared radiation, and part of the spectrum of sunlight, though invisible to our eyes. We will return to this at the end of the chapter.

The response of many perennial plants to cold temperatures is even more dramatic than the response to warming. These plants will not flower unless they are exposed to cold for a period of weeks or even months. Without the cold exposure, they cannot make FLORIGEN, and without FLORIGEN they cannot flower; the meristem just continues to produce leaves and stems. There are also some annual plants that need cold before they will flower; that's what happens with winter wheat, which is planted in the fall and then remains dormant until the end of the long spell of winter cold.

Perennial plants make a protein that switches off the *FLORIGEN* gene. That protein is FLOWERING LOCUS C or FLOWERING C (its gene is *FLOWERING C*), and we met it briefly in the previous chapter. In the presence of protein FLOWERING C, no FLORIGEN is made. The *FLOWERING C* gene itself is controlled by another protein called FRIGIDA. FRIGIDA helps to alter the chromatin histones around the *FLOWERING C* gene so that the gene is switched on. The FRIGIDA protein ensures that the *FLOWERING C* gene is copied into messenger RNA, which then directs formation of protein FLOWERING C. Once FLOWERING C protein is around, no FLORIGEN can be made and no flowers form. This series of events illustrates the kind of cascades of gene activity that govern much of what happens in both plants and animals. The switched-on *FRIGIDA* gene leads to production of FRIGIDA protein, which ensures that the *FLOWERING C* gene is on and producing FLOWERING C protein, which ensures, in turn, that the *FLORIGEN* gene is turned off; when *FRIGIDA* is ON, *FLOWERING C* is ON, *FLORIGEN* is OFF. What turns *FRIGIDA* on to begin with? That remains to be learned.

How does a plant get around this control and turn on the *FLORIGEN* gene? The answer is by turning off the *FLOWERING C* gene and losing the accumulated FLOWERING C protein. There are several ways in which this can happen. Some plants that grow from seed and flower soon after have inactivating mutations in either their *FRIGIDA* gene or *FLOWERING C* gene. In other plants, particularly perennials, the absence of FLOWERING C protein is assured when its gene is turned off after weeks of cold. Lilacs are a good example.

Usually, a plant doesn't make a lot of FLOWERING C protein if it is missing a working *FRIGIDA* gene. But there are some plants with mutant or missing *FRIGIDA* that still make a good amount of FLOWERING C protein, which keeps them from flowering very early. Such plants contain genes that regulate *FLOWERING C* gene function in ways that don't depend on *FRIGIDA*. The processes controlled by these genes are associated with stimuli within plants that are not necessarily influenced by environmental cues.

Plants and flowers that grow from underground bulbs also depend on warming temperatures to signal that it is time for their stems and leaves to come out from the dark. It seems that in these situations, a plant hormone that inhibits growth, abscisic acid, begins to accumulate in the fall. Then, after a period of real chill, that hormone breaks down while another, gibberellin, which we met in the previous chapter, increases and stimulates flowering.

Years ago, a friend of mine moved from Massachusetts to Los Angeles. She was not happy to leave her beloved lilacs behind, but she knew that lilacs require extended cold periods before flowering. Undaunted, she planted lilacs in her Los Angeles garden. Every morning in the winter months, when Los Angeles was sunny and warm, she went outside and put ice cubes around the lilac bushes. Her springtime reward was lilac blooms, not many, but enough. There are a lot of perennial plants like lilacs that only flower in spring if they have first been chilled by at least 3 weeks of cold weather. This adaptation prevents them making the mistake of flowering before it is warm enough for flowers to survive.

It's possible that what my transplanted friend was doing by chilling the soil with ice was turning the *FLOWERING C* gene down (or off). She could also have been turning down the activity of the *FRIGIDA* gene, since FRIGIDA turns *FLOWERING C* on. Turning *FRIGIDA* off requires a good, long chill, and only then can *FLOWERING C* be switched off and *FLORIGEN* be switched on.

Not all plants are like my friend's lilac bush. Many annuals, such as zinnias and nasturtium, which flower without a period of cold, don't make FRIGIDA because the *FRIGIDA* gene is either mutated or missing. There are, for example, both annual and perennial wild strains of Arabidopsis. The perennial varieties have an intact *FRIGIDA* gene and need months of cold weather before they can flower. The annual varieties have lost the ability to make FRIGIDA and need no chill before flowering; nothing holds the plant back from flowering once it is mature enough and making enough FLORIGEN.

After a long exposure to cold, the *FLOWERING C* gene is inactive. Little or none of its messenger RNA remains and its protein disappears. How does cold let the *FLOWERING C* gene escape from FRIGIDA control? The answer involves changes in the histones around which the *FLOWERING C* DNA is wound. Particular kinds of histone marks attract FRIGIDA and its partners to the DNA close to *FLOWERING C* gene and activate the gene; some of the chromatin marks that allow genes to be switched on or off were described in Chapter 5. During the long period of cold, when the *FLOWERING C* gene is slowly turned down, epigenetic eraser proteins gradually remove those marks, and epigenetic writer proteins place new marks at different positions on the histones. These changes in the positions of the methyl group decorations make it difficult for FRIGIDA to get close to the *FLOWERING C* gene and keep it switched on. The FLOWERING C protein levels decrease, and this lets the *FLORIGEN* gene switch on. Changes in the activity of several other genes also occur; one of these is called *SOC 1* and is described in the next chapter. The net result of these changes is that the plants become ready to flower when the temperature warms up again, and the light conditions and internal plant signals are right.

There are two remarkable things about this story. The first is that plant biologists managed to figure out the whole complicated set of processes. As is so often the case, the chance occurrence and observation of mutant plants that did not flower was the key. In particular, it was a surprise when some of those mutations turned out to be in genes coding for the epigenetic writer or reader or eraser proteins that affect the chemical decorations on certain histones.

The second remarkable thing is that evolution gave some flowering plants this fine level of control so they would not flower when excessive cold might destroy buds and flowers. But evolution did not take a straight path to this outcome. Why involve histones? Why not just evolve a direct way to turn down a gene's activity? In a way, the indirect path stands as a sign that evolution was at work. Biological evolution has no plan. Mutations of various kinds give rise, by chance, to new genes and new activities. If those new activities help a plant produce more offspring in its particular environment than the nonmutated parent, then those new capabilities become the norm; the parental form loses out in the competition for resources. If the new activities do not result in plants with improved reproductive success, then it's unlikely that the mutant plants will survive.

Once it was understood that the control of gene FLOWERING C is critical to successful flowering, and thus to successful reproduction, new questions about the process emerged. This is the case with most scientific discoveries; they lead to new questions. Why does the extended cold switch on eraser and writer genes that change the histone marks at nucleosomes near the FLOWERING C gene and turn it off? The answer to that question turned out to be a surprise.

The DNA of some genes, the protein genes, is copied into an RNA—messenger RNA—that contains the code for the protein. Many other genes, RNA genes, code for RNAs that function on their own in cells without coding for a protein. The FLOWERING C gene DNA codes for several RNAs in addition to the messenger RNA for the FLOWERING C protein.

One of these RNAs is called COLDAIR, as it appears to be expressed in response to cold. COLDAIR RNA is copied from parts of the FLOWERING C gene that do not code for protein. These segments are the introns that interrupt the regions that code for protein and are removed from the messenger RNA. The COLDAIR RNA sticks around on the chromatin in the vicinity of the FLOWERING C gene, where it attracts epigenetic eraser and writer enzymes that change the chemical groups decorating the nearby histones. Just how that actually works remains to be discovered. How do those enzymes find the FLOWERING C region of the chromatin? What kind of map or signal do they use to find that gene among the billions of units on the DNA? And once there, how do they distinguish one histone from another? In Chapter 5, I explained that the particular methyl distribution on histone 3 influences flowering. How do the enzymes distinguish histone 3 from the other histones in the nucleosome? Does COLDAIR help guide them? And, when they do find histone 3, how do they find the particular amino acids that need to lose or gain a methyl group so that the *FLOWERING C* gene is turned off and FLORIGEN can be made?

Part of the answer is that the eraser and writer enzymes are never alone. Like so many other proteins, they function while associating with still other proteins. Some of these combinations may help the function of writers and erasers. Others may interfere with those enzymes. It's something like an automobile assembly line. The line may be ready to work but it needs parts—tires, upholstery, and engine for example. If any part is missing, the automobile can't be built. That's the way it is with a set of enzymes, other proteins, and RNA needed to make sure a gene is on or off at the right time. If any one of the players is mutated or lost, the target gene cannot be turned on. In an automobile factory that is nothing but disaster. But in a plant biologist's laboratory, it is an important clue to the normal function of the mutated gene.

A second non-messenger RNA copied from the *FLOWERING C* gene is called COOLAIR. COOLAIR is actually a set of RNAs copied off the nonsense strand of the *FLOWERING C* gene DNA (the complementary strand that does not code for the gene); these RNAs are not messenger RNAs and do not code for protein. But, because their sequences can interact with the gene through base pairing, COOLAIR RNAs can influence the production of active messenger RNA; the interaction forms a double helix with one strand of DNA and the other of the RNA. Over the long period of cold weather, the amount of COOLAIR RNA increases, the amount of FLOWERING C messenger RNA decreases, and so the amount of FLOWERING C protein decreases. The RNA's name, COOLAIR, gives a clue as to its function. It's not yet known what makes the amount of COOLAIR RNA increase during cold weather.

As the cold weather continues, the COLDAIR and COOLAIR RNAs and perhaps other controllers of gene activation gradually turn down the activity of the *FLOWERING C* gene and make the gene unresponsive to activation by FRIGIDA. Less and less FLOWERING C messenger RNA and protein are made until, after several weeks of cold, no FLOWERING C protein is present. Turning down the *FLOWERING C* gene does not, by itself, start flowering; it just makes the plant able to produce FLORIGEN and SOC 1 proteins, both of which contribute to the initiation of flowering.

The epigenetic changes on the chromatin that turn off *FLOWERING C* gene activity are saved when cell division produces new cells to grow leaves and stems. Newly replicated DNA keeps the epigenetic marks and the *FLOWERING C* gene remains inactive through the many cell generations needed to make a flower. Yet, as flowering runs its course and new seeds are formed, the *FLOWERING C* gene is reactivated in the tiny plant embryos in the seeds. What turns *FLOWERING C* back on for the new generation? The answer to that question relates once again to changes in the decorations on the histones neighboring the gene.

The effects of winter on flowering have a name, *vernalization* (derived from the Latin for spring). There is some interesting history behind the word. It is a translation of a Russian word used by the discredited horticulturalist Trofim Lysenko. He sold his ideas about growing wheat to Josef Stalin, who adopted them as Soviet policy and elevated Lysenko to the highest status of Soviet science. Lysenko promoted the good idea of chilling wheat seedlings so that they could grow and flower promptly early in spring. But he also maintained that the effect of cold would be inherited in the next generation. This was a huge mistake, because the effect is lost in the new seeds. The result was famine in the Soviet Union.

Temperature affects flowering in ways quite apart from the intricate ways in which *FLOWERING C* activity responds to cold. One of the ways in which living things increase the sophistication and complexity of their adaptations is by the evolution of new genes. Often, this happens by the unscheduled duplication of an existing gene on the DNA, followed by the independent evolution of the two (or more) copies. The resulting family of related genes can evolve to play related but not identical roles in cells. This apparently happened in Arabidopsis with the *FLOWERING C* gene. There is a small family of related genes, one of which is called M. If M gene activity is lost, flower formation loses sensitivity to surrounding temperature. The M gene produces different messenger RNAs and proteins, depending on whether the air temperature is warm or cold. At each temperature range, a different one of its introns is removed from the initial RNA copied from the gene. The cold form of M protein turns other flowering genes off, while the warm form does not.

Still, we don't yet know how a plant senses whether its surroundings are warm or cold in the first place. Some observations may provide a clue. The levels of FLORIGEN in annual plants do go up as the external temperature warms up. So too does the amount of a regulatory protein called PIF4 that can sit right at the start of the *FLORIGEN* gene. The increasing levels of FLORIGEN track the increased amount of PIF4 at the start of the *FLORIGEN* gene. Perhaps PIF4 is working to turn up the *FLORIGEN* gene activity. Where does the PIF4 come from, and how does it know the temperature is warming? Besides associating with the start of the *FLORIGEN* gene, PIF4 can join up with a **phytochrome**, one of a group of pigments that are sensitive to red and infrared light, and hence to temperature. When they sense changes in the infrared energy, the heat-sensitive versions of phytochrome molecules undergo subtle changes in shape. It is that change in shape, influenced by temperature, which enables the phytochrome to form an association with PIF4. And this association between the heat-sensitive pigment molecule and the regulatory protein may well be what begins the cascade of molecular events that contribute to the initiation of flowering as the warmth of spring arrives.

Chapter 8

LIGHT AND DARK

Some of the differences in the flowering times of plants are brought about by temperature changes, but the daily changes in the hours of light and dark in the course of a year also play a major role. The effect of the light–dark cycle is not restricted to plants. The breeding periods for some mammals is also keyed to day length. Scientific study of this phenomenon in plants dates back only about a hundred years.

My favorite story about the influence of light on plants goes back to work in the 1920s on the origin of maize.

Botanists are intrigued by the challenge of identifying the wild plants that ancient peoples cultivated and bred for consumption. They have discovered that the same wild cabbage is the parent of broccoli, cabbage, Brussels sprouts, kale, and cauliflower. Although these vegetables look very different, they are all the same species, *Brassica oleracea*. At least one of the gene mutations that make cauliflower so different from its wild cabbage parent was discovered when an Arabidopsis mutant plant made cauliflower-like structures instead of flowers. Evidently, the external features of plants are not a big help in identifying their wild ancestors.

What wild plant is the ancestor of maize? Maize is one of those plants, like squash, tomatoes, and potatoes, that were discovered by Spanish and Portuguese explorers when they started traveling in the western hemisphere. It was likely then that the ancestor of maize was a plant indigenous to the Americas. The most obvious candidate was teosinte, a wild grass found in Central America. Serious research on this possibility began in the 1920s but immediately ran into a stumbling block. If teosinte and maize were variants of the same species, they should interbreed. That is the definition of members of a species. But botanists failed when they tried to breed teosinte with maize. In fact, the idea couldn't really be tested properly because the two plants flowered at different times. When teosinte plants produced pollen, maize had no flowers to pollinate, and vice versa. To flower, teosinte needed short summer days like those near the equator while North American maize flowers in the longer summer days of higher latitudes.

The problem of the differing flowering times was solved by manipulating day-length in greenhouses so the two plants would flower at the same time. When that was done, maize and teosinte did indeed interbreed as expected for members of the same species. Despite this evidence, many scientists clung to their own, erroneous theories about the origin of maize. Finally, by about 1980, after more than 50 years of research and heated dispute, plant biologists were convinced that the Aztecs had it right; in the Aztec language, the word teosintl actually means God's maize.

Like maize and teosinte, most plants flower in response to a characteristic ratio of the hours of light and dark. Talk of "short day" and "long day" plants is common among nursery personnel and botanists alike. The distinction depends on when FLORIGEN appears. FLORIGEN is the pot of gold at the end of the rainbow for making flowers. The rainbow itself is multifaceted: temperature, light, hormones, and other intrinsic plant processes. For temperature control, FLOWERING C is a central player. But it is an antihero; FLOWERING C needs to be turned off before FLORIGEN production can begin. In contrast, the influence of light depends on a hero, a protein called CONSTANS. CONSTANS is the central player in the effect of light on flowering. It acts directly to boost FLORIGEN gene activity, as long as that gene is no longer under the control of FLOWERING C protein.

Chapter 7 introduced the plant pigments called phytochromes, which respond to red and infrared light. These, together with cryptochromes, which respond to blue light, allow plants to sense light of different colors, and thus of different energies. The light energy causes subtle changes in the protein portions of the pigments. These changes signal to other, interacting protein molecules to initiate cascades of important effects on plant growth and flowering. Each signaling pathway, and the cascade that follows, depends on the control of a set of genes and the RNAs and proteins those genes specify.

The day–night cycle is key to flowering. Some plants flower in fall when days are growing shorter and nights longer: think about chrysanthemums and toad lilies. Some, such as petunias and impatiens, flower in late spring or summer when days are as long as 16 hours, depending on latitude. One of the camellia bushes in my garden produces red blossoms in early spring, while another variety bears pink flowers in October.

Arabidopsis too has variants that flower when days are short and others that flower when days are long. The timing reflects the presence or absence of mutations in particular genes. It all comes down to how plants sense light and how varying periods of darkness and daylight influence the activities of FLORIGEN and other genes important for starting flower formation.

Light influences flowering time in ways that depend on seemingly unrelated genes in the plant's DNA. These genes code for the proteins that construct what is known as the circadian clock, which drives the circadian rhythms that influence the behavior of organisms, most obviously the sleep–wake cycle. As far as we know, all animals and plants have such an internal clock. These clocks evolved early in evolution and are roughly tuned to Earth's twenty-four-hour, night–day, dark–light cycle. Well before green plants appeared on Earth, genes for a circadian clock were present in fungi. Humans also have circadian clocks. If you have ever taken a long plane ride through many time zones, you know what a major effect your clock can have on how you feel. That's the human circadian clock at work. Describing how the clock works, in animals or in plants, would need a whole book of its own. The important thing is that many genes are switched on and off in tune with the clock, including flowering genes.

The Arabidopsis plants that many botanists study like to flower when days are long and there are as many as sixteen hours of daylight; that occurs around the summer solstice. Several genes in the circadian clock network influence this response to light and lead to the production of FLORIGEN during long days. Some of these genes code for the manufacture of the phytochrome and cryptochrome pigments that sense light of particular wavelengths. Once the light is sensed, a cascade of genes is turned on; each active gene in the cascade assures that the next gene in the path is activated. At the end of this cascade is the gene called CONSTANS. Some of the genes in the cascade code for proteins that can keep CONSTANS from working, just as FLOWERING C keeps the FLORIGEN gene from being switched on. Other genes in the cascade code for proteins that allow the CONSTANS gene to be switched on. And the

Plate 1: Reproductive structures within a flower. Stamens carrying pollen surround the central carpel, which ends in the stigma.

Plate 2: A crocus with veined, coloured sepals, like petals (left) compared with typical green sepals, here in a sunflower (right).

Plate 3: Hummingbird with its long beak in a red monkey flower
(*Mimulus cardinalis*).

Plate 4: The extended petal of a bee orchid (*Ophrys apifera*) mimics the
appearance of a bee, attracting male bees as pollinators.

Plate 5: (left) Butter-and-eggs plant (*Linaria vulgaris*), and (right) close-up of its relative, the snapdragon (*Antirrhinum*). The petals of these flowers are fused to form a tube.

Plate 6: Close-up of portion of an oxeye daisy (*Leucanthemum vulgare*), a compound flower, showing the central disc flowers, surrounded by white ray flowers.

Plate 7: Normal sunflower (*Helianthus*) with ray flowers surrounding a head of dense disc flowers (left), and a mutant form with almost entirely ray flowers (right).

Plate 8: The two yellow markings of the monkey flower (*Mimulus lewisii*) guide bees into the flower.

Plate 9: White azalea with pink streaks and spots. Such random coloured markings can result from jumping genes that block the production of anthocyanins in some parts of the petal but not in others.

proteins produced from some genes are important for making the CONSTANS protein stable through the long days. In this way, signals from particular phytochromes and cryptochromes contribute to the accumulation of CONSTANS protein. Once made, the CONSTANS protein binds to the control regions of the *FLORIGEN* gene and turns up the activity of that gene as long as the FLOWERING C protein has been turned off.

The *CONSTANS* gene is switched on and off, respectively, by several proteins that are made and degraded in sync with the twenty-four-hour clock and the day–night cycle. The amounts of the several proteins that control *CONSTANS* gene activity are also affected directly by the amount of light the plant receives, which means the lengths of the night and day. Because of this, *CONSTANS* activity varies between long and short days. During long days, the accumulation of CONSTANS messenger RNA and protein begins by mid-day. By dusk, it is at its maximum and the amount of CONSTANS protein then falls dramatically. By then, however, the amount of FLORIGEN in the cell has increased substantially so flowers can be made (assuming that no *FLOWERING C* is being made). As night begins after the long day, the CONSTANS protein is degraded and is kept low by signals coming from another of the phytochromes, even though the messenger RNA from the *CONSTANS* gene remains high until night is almost over. CONSTANS protein remains low during much of the night. Consequently, the activity of the *FLORIGEN* gene begins to fall. When dawn comes again, the cycle begins anew.

When days are short, it's already getting dark when the genes needed to ensure production of CONSTANS are switched on; only small amounts of CONSTANS accumulate. Other mechanisms also contribute to the loss of FLORIGEN. Nevertheless, some plants do flower in short days. Not much has been done to study the mechanism of flowering induction in short days but there are hints that the hormone gibberellin promotes short-day flowering.

The *CONSTANS* gene appears to have been around since early in the evolution of plants. Similar genes occur even in unicellular algae, the ancestor of all land plants. The activation of the *CONSTANS* gene in algae also depends on the working of the biological clock.

We've now looked at how internal signals of maturity, and the sensing of warmth and light lead to gene cascades that end at the *FLORIGEN* gene. One or another of the internal or environmental cues may be predominant in different plants or under variable environmental conditions. The upshot of all this complexity is that FLOWERING C is diminished or gone, the *FLORIGEN* gene is turned on, and FLORIGEN protein can travel through the stem to the meristem. Our story now moves to the growing tip.

PART IV

Shaping a Flower

Together, maturation, temperature, and light influence when a plant will begin to flower. In most plants, production of FLORIGEN marks the time to flower. However, in the process of evolution, some plants developed the ability to flower and make seeds without active FLORIGEN. Other plants flourished by sending out roots that give rise to new plants. The survival of new plant properties which interact successfully with particular environments is the process of natural selection. Natural selection is a key mechanism in evolution, and it can lead, over time, to complex and varied adaptations to a particular environment—different solutions to the problem of survival. In this case, the outcomes include plants that flower in the absence of FLORIGEN, and plants that don't flower at all yet still manage to reproduce given a hospitable environment.

Whether initiation of flowering is dependent on FLORIGEN and/or on other signals, it results in changes in the structure and chemistry at the growing tip, or meristem. These changes start a cascade of events. The cascade reflects the switching on and off of particular genes and it evolved to result in the formation of that wonderful adaptation, a flower. This is a huge and critical task for the plant.

Chapter 9 traces the story of the switch to flower formation. The meristem becomes a construction site that produces a complex structure. Because flowers have such different shapes, events at the meristem can be quite different depending on the kind of plant. The description in Chapter 9 is necessarily general. The simplest shape is made up of a circle of green sepals surrounding a circle of simple petals and within that circle, the stamens and a carpel. Wild roses are good examples of this kind of flower. But many, probably most of the flowers we see and admire are not like that. Gene mutations have provided much of the variation seen in wild and cultivated flowers.

In addition to the large variety of overall flower shapes, the individual floral organs have distinct shapes in different flowers and sometimes even within a single flower. In some flowers, the stamens and carpels are fused into a single structure. Floral looks can be deceiving. Chapter 10 describes what has been learned about the complex processes by which flowers with special shapes such as snapdragons and daisies form.

Chapter 9

THE CONSTRUCTION SITE

What happens when a plant is ready to flower? It's one thing to keep on growing longer stems with more leaves. It's an entirely different task to set about making a flower. Yet, the two different tasks occur at the same place, the meristem. How does a meristem switch from leaves to flowers?

To our eyes, the meristem looks like a tiny point. But through a microscope, a unique landscape is revealed. The more or less round meristem structure contains several layers of cells and is somewhat thicker at the center than at the periphery, depending on the plant. This architecture makes the meristem look like a dome.

The mound of cells at the very center of the meristem are stem cells, cells capable of changing their programs to become the various kinds of cells that make a flower. A protein called WUS is produced in one of the lower layers of cells. The job of the WUS protein is to make sure that the cells in the overlying layers remain good stem cells. Like almost all of the genes and proteins mentioned in this chapter, WUS protein regulates the activity of other genes by switching them on or off.

Plant meristem stem cells can divide to produce one daughter cell that is another stem cell and one cell that is now reprogrammed to begin the building of a flower. To begin with, there are probably very few such cells in the meristem but they divide many times and their reprogrammed daughters are pushed out of the center to the periphery of the meristem. At the periphery, they form concentric circles of cells surrounding the central mound of stem cells. Seen in a microscope, the meristem begins to look like a hat with a wide brim. The cells in the brim are those that have been reprogrammed to give rise to a flower. One distinguishing mark of these cells is that they contain the plant hormones auxin and gibberellin, which help ensure that they will develop into a flower. These hormones remain important through the whole process. A gene called *APETALA I* is also active in these cells, although at low levels.

Some cells around the center of the meristem produce especially high levels of auxin. These cells grow and multiply more rapidly than their neighbors and form layers within the several concentric rings. Each cell has what can be thought of as having an address defined by its location in a particular concentric circle and a particular layer of cells. The individual addresses are associated with the activity of particular genes and thus the cells at different addresses become specialized to build each of the several flower parts.

No matter whether the plant is one that flowers in long days or short days, or in spring or fall, or is an annual or a perennial, the signal is usually the same: the protein called FLORIGEN. Sometimes, flowering can occur in the absence of FLORIGEN if there is enough gibberellin around. But as a general rule, FLORIGEN has the central stage for the initiation of flowering.

A signal alone does not, of course, mean much. Something needs to recognize the signal and react. In the meristem cells prepared to form a flower, that "something" is two additional proteins. When FLORIGEN arrives, the three proteins join up with one another and the protein complex starts things off. The partnership between the three proteins is the key. It will be no surprise by now that what that key does is switch on another set of genes that are themselves switches that operate on particular genes. This kind of chain of events is the way many developmental processes in plants and animals are set in motion; different genes that code for protein switches are activated in turn by the protein(s) produced by a previous gene in the chain. Finally, the few genes that will actually carry out the task at hand get switched on. One of these, *SOC1*, acts, as does gibberellin hormone, to turn up *LEAFY*. And there are other genes and proteins too that influence *LEAFY* gene activity. Thus, many genes, some with redundant purposes, can ensure construction of a flower. If you are a flowering plant, making a flower really matters; without it there will be no offspring. So evolution has adopted several ways to get the job done.

LEAFY and *SOC1* are the culmination of the integration of the various environmental and internal signals that promote flowering. The SOC1 protein activity also tends to increase *LEAFY* activity. And what does the LEAFY protein do? It coordinates the activity of a whole group of genes whose gene products are important for actually making the flower. In a sense, LEAFY makes it all happen.

One of the genes turned up by LEAFY is *APETALA 1*; its name signifies that if it is mutated or lost, no petals form on the flower. The LEAFY and APETALA 1 proteins turn off genes that stop flower formation and turn on the genes needed to make the flower. Beside all the other things that control flower formation that were described in the previous chapters, there are indeed genes that block flower formation even at this stage. One of them stops flowering in many plants beside Arabidopsis (examples include snapdragons and rye grass). This gene (*TFL1, terminal flower 1*) antagonizes

LEAFY and APETALA 1 and must be turned off if the meristem is to become a productive flower construction site. But LEAFY and APETALA 1 have their own tricks. They associate with the region near the end of the gene on the DNA and that's enough to turn *TFL1* off. Once LEAFY and APETALA 1 are in control, flower formation begins in earnest and cells at the periphery of the meristem multiply rapidly.

The switched-on meristem will not look very different to our eyes, but the changes can be seen with powerful microscopes. The formation of secondary bumps as concentric circles around the center of the meristem is a consequence of aspects of continuing cell division. Cells multiply at different rates. They also divide in different directions such as crosswise or longitudinally or vertically. And the new cells can begin to acquire distinctive shapes as they join one or the other of four concentric whorls.

Each of the four whorls will produce one of the four organs of a flower. Sepals form from the outermost whorl of cells. The next whorl toward the center will give rise to the petals. The third whorl from the outside is where the stamens will arise, and the fourth, the innermost whorl, will form the female organ, the carpel or pistil, where the egg will be found. Slower rates of cell division define the borders between the four whorls. The slower rate of cell division depends on the action of other genes. A gene called *SUPERMAN* is turned on between the third and fourth whorls, the ones that make stamens and those that make the carpel. In the presence of the SUPERMAN protein, the cells grow more slowly and maintain the boundary. Another, similar protein called RABBIT EARS does the same thing in the boundary between the second and third whorls, the ones that form petals and stamens. That, of course, doesn't explain how it happens!

How does the plant, or the meristem, know that cell division must slow down in between the whorls? What tells the cells between the whorls to slow down? For that matter, how does the meristem know to form four, not three or five, whorls? We do not yet know. Again, we see how in science the answer to one puzzle raises many new ones.

Soon after the bumps of new cells appear around the edge of the meristem, LEAFY and APETALA 1 help switch on another group of genes called *SEP*. Up until this point, the cells are programmed for flowering but not yet actually making a flower. Once the *SEP* genes are switched on, their products provide the "no-return" signal for flower formation.

Incidentally, the *APETALA 1* gene has what we might call a cousin. Families of genes that produce similar, sometimes almost identical proteins are common in plants and animals, and have usually arisen, as we noted earlier, from gene duplication events in their evolutionary past. Sometimes, such nearly identical proteins may fill in for one another. *APETALA 1* is part of such a gene family and its action is (partly) redundant to another gene called *CAULIFLOWER*. A mutation in *CAULIFLOWER* has no obvious effect on an Arabidopsis plant. But if both *APETALA 1* and *CAULIFLOWER* are knocked out, the meristem looks like a tiny cauliflower, and is made up of many different flower meristems. That's in fact what a cauliflower is: an enlarged white meristem of a mutant cabbage plant that has trouble developing flowers.

To return to our story, we now have four concentric rings of cells being formed at the meristem and the *SEP* genes have been switched on, preliminary to actual flower formation. The four whorls are regions where particular genes are turned on or off, genes that instruct the formation of one or the other of the four typical organs. This general scheme was discovered more than twenty years ago through the study of Arabidopsis. Since then, many details, although not all, have been filled in: some by elegant microscopy and others by the use of genetics. Some details were illuminated by studying flowers that don't conform to the common floral structure of sepals, petals, stamens, and carpels. The scheme can be thought of as four unique genetic programs, one for each whorl, and therefore one for each of the usual four flower parts. But bear in mind that the neat outline of flowering described here hides a lot of variation and complexity among different plants and even within a species.

A few genes are active in all four whorls including LEAFY, APETALA 2 (another member of the APETALA gene family), and one or more of the SEP genes. In addition, there are special organ-specific programs. The programs for formation of each of the four organs are defined by only four genes. These genes are APETALA 1, PISTILLATA, APETALA 3, and AGAMOUS. These four genes work in different combinations to define the four floral organs. APETALA 1 is associated with the outer parts of a flower (sepals and petals) while AGAMOUS is involved with the formation of the reproductive parts of the flower.

The formation of sepals in the outermost whorl depends on the continuing activity of the APETALA 1 and APETALA 2 genes. In the next whorl, where petals form, APETALA 1 stays on and APETALA 3 and PISTILLATA are activated. In the adjacent whorl, where stamens are formed, APETALA 3 and PISTILLATA remain on while APETALA 1 is turned off, APETALA 2 is turned down, and AGAMOUS is turned on. These genes switch themselves off or on; other genes including RNA genes interact with the gene sequences to modulate their activity. After a couple of days the accumulation

of AGAMOUS protein leads to the switching on of a gene called
KNUCKLES, which in turn alters the way that histone 3 is methylated around the WUS gene and switches that gene off. WUS protein is no longer needed to stimulate formation of new stem cells;
the flower is by now well on its way to being fully formed.

The only task still to be accomplished is the formation of the
carpel in the fourth whorl. That requires AGAMOUS and depends
on switching off APETALA 3 and PISTILLATA which would, if still
active, stop carpel formation.

Just five genes (APETALA 1, 2, and 3, PISTILLATA, and AGAMOUS),
sorted in different ways, direct the formation of the whole, complex, flower. The products of these genes affect the activity of other
genes that regulate the genes needed to actually build the flower.
Their target sequence on the genome appears to be merely ten
bases long: CC followed by six As or Ts and then GG. These same
proteins can also influence the boundaries that define the different
whorls and keep them independent of one another. For example,
the activity of APETALA 1 or its gene product, APETALA 1 protein,
does not allow AGAMOUS to be active and the reverse is also true:
APETALA 1 must be switched off before AGAMOUS can be activated. None of the gene products of these special genes act alone;
all of them join forces with other proteins and different molecules
to carry out the development program.

Think of it like a symphony in four movements. The entire symphony depends on the orchestra and conductor: APETALA 2 and
SEP. The first movement depends on one theme, APETALA 1. The
second movement builds on the first by adding APETALA 3 and PIS-
TILLATA. No hint of the APETALA 1 theme is left in the third movement, which introduces the AGAMOUS theme. Finally, AGAMOUS
alone completes the symphony with the fourth movement.

Many questions remain. For example, AGAMOUS protein helps to turn *APETALA 1* off and promotes formation of stamens and carpels, but what turns off *APETALA 3* and *PISTILLATA* in the fourth whorl where the ovary and other female organs form? At this time, we do not have an answer.

Variations occur when one of these genes is mutated, or missing, or is turned on in the wrong place because of mutations elsewhere in the genome. That's what happens in crocuses. *PISTILLATA* and *APETALA 3* are expressed in the outer whorl so instead of sepals, the plant makes extra petals.

Perhaps it's surprising that plant scientists have spent time and money studying crocuses. Why crocuses? Because crocuses are big business. The stigmas on the purple crocus (*Crocus sativa*) carpels are the source of saffron, and saffron costs thousands of dollars a pound, depending on its country of origin. Fortunately, only a few threads are needed to flavor a dish.

Boundaries are also important within a whorl. Boundaries ensure the formation of multiple sepals and petals in the first and second whorls, respectively. The stamens and carpels are themselves constructed of several distinctive parts, including the pollen and eggs, so boundaries matter there as well. Various genes that influence boundary formation and maintenance in these developing organs have been identified. So far, however, they don't shape up into a straightforward story. Some of them seem to slow cell proliferation or size or dictate particular shapes so as to make a border. But essentially all of the genes that have been identified are genes that turn other genes on or off. Again, many questions remain.

The real actors, the genes that give a flower green sepals, or shiny, shaped petals, are just beginning to be studied. Experiments suggest that at least twenty-three separate genes may be involved in determining the size and shape of Arabidopsis petals. It will be interesting to learn how these genes and the proteins or RNA that they produce exert their effects. Some part of size and shape will be governed by the differing rates at which cells multiply at different positions in the petal.

Although much of what we know about how flowers form comes from studies with Arabidopsis, work with several other plants confirms the generality of the mechanisms. Important agricultural crops like maize and commercial nursery plants like petunias and snapdragons (which predate Arabidopsis as an experimental model) are also laboratory subjects. Much of what is learned with each organism is readily applicable to others, but it's not always simple to detect the similarities in scientific publications because the same gene often has a different name in each organism. Still, the relationships are there.

So far, it seems from this chapter that all flowers can be described as variants of an arrangement of sepal, petals, stigmas, and carpels. But obviously, many flowers, including quite common ones, are not so simple. How, for instance, do genes account for all the daisy-like flowers like gerbera, asters, sunflowers, and daisies themselves? The next chapter will look into these unusual forms.

Chapter 10

SPECIAL SHAPES

Chapter 9 describes the simplest flower that can emerge at a meristem. Such blooms, like wild roses, are radially symmetric. But many flowers are not so simple. Daisies appear to have a radially symmetric array of petals surrounding a center made of small structures. Some, like orchids, snapdragons, and foxglove blossoms, are bilaterally symmetric rather than radially symmetric: they have a line of symmetry that runs top to bottom. Such bilateral symmetry is familiar because it is the way human bodies are organized from the top of the head to the feet. The evolution of different flower shapes went hand-in-hand with the evolution of pollinating insects and birds, to maintain efficient pollination of the plant and provision of nectar to the pollinator. Each shape appears to make a hospitable entry for a particular pollinator (Plate 3).

There are over 26,000 species in the orchid family of plants, the Orchidaceae. At first glance, the wild orchids don't appear to be similar to the large, elegant orchids available for sale. Orchid breeding by amateur enthusiasts is a popular pastime. It is also of great commercial interest. A result of all this popularity has been active and well-funded research on this varied family of flowers. Consequently, we have substantial insight into why orchids are the way they are. Even though they look so different from a radially symmetrical flower, the development of orchids and other bilaterally symmetrical flowers depends on the same four whorls and the same genes as does Arabidopsis. Particular mutations give rise to the variant shapes. Orchids, like crocuses, have no sepals because a mutation causes the *PISTILLATA* and *APETALA 3* genes to be activated in the outside whorl of developing orchids where they would be silent in simpler flowers.

Also, orchid petals are not all identical. There is an unusual petal, the "lip" petal, at the bottom. The presence of multiple, variant copies of *APETALA 3* may influence the special shape of that lip. For some orchids, the lip is an important platform from which insects enter the blossom looking for food and consequently pick up pollen (Plate 4). Insects are not the only animals that serve as orchid pollinators. Different orchids use butterflies, birds, and even small mammals to carry their pollen to other blossoms. Orchids have several other unusual features. The carpel and stamens, for example, are fused into one structure, and the genetics and epigenetics behind this variation are still being explored.

Snapdragons (*Antirrhinum majus*), another plant with bilaterally symmetric flowers, have been a favorite tool of plant biologists for a very long time. Most of the genes that control the formation of flowers in Arabidopsis have cousins if not twins in the snapdragon genome.

One of the first wild flowers I learned to identify, butter-and-eggs or toadflax (*Linaria vulgaris*), is a close relative of the snapdragon, and one in which the bilateral symmetry is easy to spot because one of the petals is bright orange while the rest of the flower is yellow (Plate 5a). Even at a quick glance, the petals of a snapdragon or butter-and-eggs flower look quite different from a radially symmetric flower like a wild rose. There are two top petals, two side petals, and a bottom petal. In the eighteenth century, Linnaeus himself observed unusual butter-and-eggs plants with radially symmetrical flowers but knowing nothing about genes, he could not understand why this occurred.

At present, more is understood about how bilateral symmetry develops in snapdragons than in butter-and-eggs or other flowers. Serious research on snapdragons began about a century ago. Early on, mutants that had lost the bilateral symmetry were discovered. In fact, snapdragon mutants occur quite frequently and they often return to the original radially symmetric form (the mutation is not stable). This behavior is the result of an active group of moveable DNA elements in the genome. They can move into a gene causing a mutation and then pop out again, restoring the original gene activity.

Like so much else learned so far about the development of flower form and shape, most of the snapdragon experiments have identified regulatory genes that switch other genes on or off. These genes function in the development of petals (whorl 2) and stamens (whorl 3) and establish a top and a bottom for the developing flower. Rather than maintain their individual identity, the five petals of snapdragons are fused into a single tube (Plate 5b).

One key gene that is turned on in the developing snapdragon is called CYCLOIDEA. Another gene, DICHOTOMA, is also important for the asymmetry in the final flower. These two genes are so closely related to one another that they are likely to have originated by duplication of an ancestral gene. If both these genes are mutated or missing, the flowers will be radially symmetric. CYCLOIDEA is a good example of a gene subject to epigenetic control; it is turned off when its DNA is methylated. Like all the other genes and gene products described so far in this book, the proteins coded by CYCLOIDEA and DICHOTOMA influence the form of a flower by modulating the activity of other genes. They appear to do this by their ability to bind to particular DNA sequences and thereby activate a neighboring gene. We can begin to understand how genes like *CYCLOIDEA* might work from the fact that from the earliest development in the meristem, they are particularly active in the region that will form the top petals of the flower. Also, the cells near the tip of an emerging petal and those near the bottom grow and divide at different rates and acquire different shapes. In complicated and sometimes antagonistic ways, other genes also influence the differential development of top and bottom petals and together impose the asymmetry.

While some radially symmetric flowers don't contain a *CYCLOIDEA* gene at all, others do have the gene. The Arabidopsis genome includes a *CYCLOIDEA* gene although its flower has a symmetrical array of four white petals. In such instances, the localization and level of the gene's activity is different from that in plants like snapdragons. Both *CYCLOIDEA* and *DICHOTOMA* and a few related genes also occur in flowers with quite different symmetries. Again, the different symmetries reflect the position in the developing flower where the genes are active. We have yet to learn what accounts for the uneven distributions of gene activity in different parts of the developing flower organs.

Snapdragons are a good example of how the success of particular flower shapes in nature has depended on the coevolution of successful pollinating agents such as birds or insects. Bumblebees are efficient pollinators of snapdragons. The evolution of the flowers has yielded an architecture that is particularly suitable for bumblebees. And the asymmetry is not just in the way the petals are arranged around the stamens and carpel but in the shape of the petals themselves, the differences between the two sides of the petal, and the shape of the stamens. It's worthwhile looking more carefully at flowers than most of us usually do; there are all kinds of distinct patterns to see.

At first sight, gerbera, daisies, sunflowers, and similar blossoms seem to be made up of an outer circle of only petals and a central disk of tiny bumps. While the whole blossom may appear to be radially symmetrical, the reality is a lot more complicated. In fact both the outer circle and the inner disk are collections of individual flowers. The petals in the outer circle are each a bilaterally symmetrical flower (like snapdragons), called a **ray flower**, and the central disk contains many tiny, radially symmetrical **disc flowers** (Plate 6). It's not easy to see the ray flowers' parts except for the petal itself; the rest of the parts are close to the place where the ray flower joins the disc. The disc flowers are not very easy to see either because they are so small; a good magnifying glass can help, but not always satisfactorily.

This architecture explains why flowers like daisies are called compound flowers. What happens during early flower formation to construct two rather than one kind of flower? Are there eight whorls instead of four: four for the petal flowers and four for the center flowers? Or is the development of these compound flowers quite different? We don't yet know. The story of flower formation is still incomplete.

We do know that reproduction in a plant with compound flowers is a cooperative effort between the two kinds of flowers. The ray flowers are specialized for attracting the pollinators essential for seed production but are themselves sterile. In contrast, the disc flowers have fertile pollen and fertile eggs and depend on the pollinators, usually bees, attracted by the ray flowers, for fertilization. A large sunflower blossom can produce one thousand or more seeds.

The bilateral symmetry of the ray flowers in sunflowers depends, as might have been suspected, on that master of asymmetry, a *CYCLOIDEA* gene. The gene is not expressed much if at all in the radially symmetric disk flowers. The bilaterally symmetric ray flowers of sunflowers disappear in plants that contain mutations in a member of the *CYCLOIDEA* gene family; they become like the radially symmetric disc flowers. Different mutations in members of the *CYCLOIDEA* gene family make for even stranger results. In plants with one such mutation, the disc flowers are replaced with ray flowers. This mutation does not change the protein coded for by the gene but instead switches it on in places where it is usually silent. In this way, what should be radially symmetric disc flowers turn out to be bilaterally symmetric ray flowers (Plate 7). The scientists who discovered this realized that the mutant flowers had already been discovered and described by the nineteenth-century painter Vincent Van Gogh. Van Gogh's many sunflower paintings are famous and illustrate flowers with the normal distribution of ray and disc flowers as well as those where the disc flowers are lost and replaced with ray flowers.

The arrangement of disc flowers in a sunflower head is extraordinary. They grow in graceful spirals from the center in clockwise and anticlockwise directions. And the number of spirals in each direction is a Fibonacci number, part of the Fibonacci series, obtained by taking the sum of the two previous numbers to make the next member of the series (0, 1, 1, 2, 3, 5, 8 …). Long before their discovery by the thirteenth-century mathematician Leonardo Pisano, evolution was using them in a variety of ways. For example, the starfish and all their relatives in the phylum Echinodermata have five arms, while wild roses have five petals, and lilies have 3, both being Fibonacci numbers. The characteristic numbers of petals (not including sepals that look like petals) on most plants are Fibonacci numbers; three and five are common, and only rarely does a wild flower have four petals. Many spiral arrangements in nature, be they in snail shell curves or the array of seeds in the center of compound flowers like sunflowers, are Fibonacci numbers. What's more, the number of clockwise and counterclockwise spirals made up by the disc flowers in the sunflower head make up neighboring Fibonacci numbers: 34 and 55, or 55 and 89, for example. The generally agreed upon explanation for why evolution would favor such numbers is that the arrangements allow for the most efficient packing in the disc.

So far, this chapter has been concerned with how the overall shapes of different flowers come about. The shapes of the four individual floral organs on different plants can also be distinctive. Some of the genes involved in these processes have been identified and in time it will be possible to summarize the way they function. For now, however, it's only possible to outline the kinds of questions that scientists are trying to answer. Questions about petal formation illustrate the thinking.

We have already touched on how petals separate. The formation of boundaries between petals is related to the formation of boundaries between the four whorls that define the flower itself. As in the case of the whorls, boundary formation is associated with a decrease in the rates at which cell division and cell growth occur. But what of petal shape?

The petals on radially symmetrical flowers are generally all alike. Rose petals are usually rounded, for example, while those on lilies are often pointed. The petals on bilaterally symmetrical flowers are sometimes closely similar but in other cases, orchids for instance, the several petals may have distinctive shapes. Some petals sport ruffles around the edges while others are smooth throughout. How do such variations arise? We touched on part of the answer to this question earlier, when we considered the asymmetry of the top and bottom petals of snapdragons. For any particular flower, only the beginnings of answers are available so far, but there are some general concepts that form the basis of the changes.

Imagine a petal (or a leaf for that matter) forming, growing, and taking shape. Either or both of two basic processes might be at work. One is growth by the formation of new cells by division of the existing cells. The other is the growth of individual cells—how small or large they are. Growth rates and the direction of growth can vary, as can the rate and direction of new cell formation. Differential rates and/or direction of one or both of these processes will lead to different petal shapes. The growing petal may become longer or wider or more circular and the petal can change shape as it matures. In many flowers, like roses, petals need to unfold from a compact bud. In others, the petals first appear as tubes that unroll. Finally, signals are needed to stop the growth of petals when they have reached the appropriate size; the ray flowers of a daisy are a lot smaller than those of a sunflower. Figuring out how all that works is challenging especially because several different genes are involved at each stage.

One way to approach the study of petal formation is to divide the process into a series of steps, collect petals at each step, and ask what genes are active at each step. The cells in the petal will contain RNA from those genes that are turned on and those RNAs can be isolated from the petals and identified. Depending on the plant, it is possible in this way to notice that at different stages of petal development, the active genes may foster new cell formation or expansion of existing cells. And it's not surprising that although different plants depend on similar genes to control petal shape, the activity of those genes differs in timing and location.

So far, I have described what we know of the formation of a flower, and its general shape. But in order to draw the attention of pollinators, flowers must do more. They must decorate themselves with color and scent. These form the subject of the final part of the book.

PART V

Decorating a Flower

Much as we might like to think that flower colors and perfumes are there to please us, it's really the bugs and birds that matter. Bees, butterflies, birds, and flowers evolved together in a mutually beneficial way. Flower color and perfume entice these animals to browse. Poking into the blossoms for nectar, the animals pick up pollen from stamens and deliver it to nearby or distant carpels for pollination. The flower invests a lot of chemistry and energy in making the decoys and the nectar. In return, the attracted insects and birds oblige by ensuring pollination and thus successful seed production. Without the animals, there would be little pollination, no seeds, and no future flowers or fruit. That is why fruit growers worry about the decreasing bee population. Without bees there will be few peaches, almonds, squash, or pumpkins.

Fortunately for understanding flower color and perfume (and many other aspects of flowers), interested amateurs and scientists have described natural mutant variants and collected their seeds for many years. The seed collections are called germ plasm collections. Carefully annotated collections of seeds exist in many botanical gardens and arboretums around the world. Catalogs, now accessible through the Internet, allow plant scientists, practical breeders, and amateur gardeners to learn what is available and obtain access.

The tool chests for painting and perfuming flower petals begin with genes. In trying to understand flowering, scientists studied chance mutations, like that of LEAFY, and then tried to discover what the protein or RNA products of the normal form of the mutated genes do in the plant. In contrast, when it comes to color and perfume, a lot is known about the functions of the RNAs and proteins the genes produce. Many are directly engaged in the manufacture of the decorative molecules. They are not all genes that switch other genes on or off.

Extensive knowledge about plant production of color and perfume has accumulated over the centuries, as humans sought to use flower colors (Chapter 11) and perfumes (Chapter 12) to enhance their physical attractiveness, decorate their clothes, and flavor their food. The production of these molecules was an ancient industry. Then, during the nineteenth century, large-scale commercial production of dyes and perfumes spawned huge corporations that stimulated research on these substances.

Preparing dyes and perfumes from flower petals is an expensive operation and depends on unreliable elements like weather. Industrial laboratories were therefore established to determine the chemical structures of the dyes and odors isolated from plants. The laboratories developed synthetic methods to produce these chemicals more cheaply and reliably than by isolating them from natural products. This effort could be said to represent the birth of organic chemistry.

The effort to make synthetic dyes has been quite successful. In part, this is because the spectrum of the natural and synthetic dyes can be compared to indicate how close the chemists have come to reproducing nature. The story is different with perfumes. Until recently, there were no objective measures to compare the natural odor of, for example, a lilac with the synthetic product sold commercially as a "lilac" fragrance. Also, most floral perfumes are mixtures of different aromatic compounds and some of these mixtures are quite complex. In the past, manufacturers had to rely on subjective measures in trying to reproduce natural odors. That has changed with the introduction of modern analytical chemistry techniques known as gas chromatography and mass spectrometry that can identify complex molecules in a mixture.

The success of the organic chemists in identifying the dye and perfume molecules was an important element in the identification of the genes responsible for the manufacture of the molecules in plants.

Chapter 11

PAINTING THE PETALS

"I think it pisses God off if you walk by the color purple in a field somewhere and don't notice it."

Alice Walker, The Color Purple

The color of flowers catches our eyes whether or not we are looking for it. A single bright yellow dandelion growing from a crack in a sidewalk invites city dwellers to look, and children to pluck. In late summer, blue chicory in an abandoned empty lot brightens an unlikely spot. A mass of multicolored tulips in a garden dazzles. Living in Washington, DC as I do, I silently thank Lady Bird Johnson every spring for the glorious flowerbeds she introduced to the city. And on an odd-shaped empty plot bordering one of our city's many challenging traffic circles, volunteers have planted and tend a brilliant array of flowering plants year after year, season after season.

Flower color is also a significant economic enterprise. According to one estimate, more than $70 billion is at stake. Companies invest a lot in interbreeding and genetic engineering to make new colors and we, the public, eagerly buy novel varieties of flowering plants at high prices. The quest for a blue rose is a case in point. That quest has been going on among plant breeders for more than a century. Once genetic engineering techniques became available, they were quickly applied to the development of a blue rose. I found announcements of success in 2004, 2008, and again in 2011. In fact, the roses pictured look more like a light purple (mauve) than blue. It's a step in the right direction, but there is a long way to go and success is likely to yield a rose that costs twenty to thirty dollars a stem. Some think they have seen a rose of true blue hue but, sad to say, it's probably a white rose dyed blue; that's not uncommon.

On the other hand, breeders have been successful in developing and breeding blue petunias. That's not because they altered genes so that blue pigment was made. Rather, the manipulated gene works to sustain the normal acidity of the petals. After the alteration, the acidity of the petals decreases and the flowers become blue rather than red.

People with gardens know that red flowers often attract butterflies and hummingbirds. In one of Costa Rica's splendid national parks, I saw a vine with bright red blossoms that is a member of the same group as azaleas, blueberries, and heather; no common name was available from our naturalist guide who reported that it was in a group known as either *Ericaceae* or *Cavendishia*. Its bright red blossoms looked like long, upside-down tulips, hanging in a row from a long stem. The guide explained that this arrangement made it easier for hummingbirds that reach inside the flowers searching for nectar. As the birds do so, they pick up pollen on their feathers and then deposit the pollen on carpels to fertilize eggs. Then came the "truth is greater than fiction" fact. After pollination, the blooms rotate so that their vessel-like flowers are turned up. Hummingbirds no longer visit but it is easier for tanagers to reach the fruit and seeds thereby promoting dispersal. I've never been sure whether to believe that story. But it is well known that some flowers change their color and shape after being pollinated.

The genes responsible for making flower pigments code for protein enzymes that enable the chemical reactions required to give flowers their colors. Mutations in those genes, as well as in genes that code for molecules that regulate them, produce the diversity of color and pattern we see. Interest in the genetics of flower color began long before the mid-twentieth-century adoption of Arabidopsis as a model plant. Most of what is known about the genes responsible for flower color comes from research on snapdragons, petunias, and morning glories.

The story of flower color focuses on two very large, but very different, molecules: protein enzymes and dyes (pigments). Like all protein molecules, proteins that operate as enzymes are huge macromolecules. They may have, beside thousands of carbon, hydrogen, and oxygen atoms, other atoms such as nitrogen, sulfur, and phosphorus. The structure of each such enzyme is determined by one or more genes in a plant's DNA. Information flows from the genes to messenger RNAs that are then translated, through the genetic code, into the protein enzymes, that then catalyze the chemical reactions leading to the pigment molecules. Other genes and corresponding proteins and RNAs determine when, where, and how much of the flower color will develop.

The colored dye molecules are very different from proteins. Although large compared to water and sugar molecules, they are not nearly as huge and complicated as proteins. Some of the pigments have as many as a couple of hundred atoms, mainly carbon, hydrogen, and oxygen unaccompanied by nitrogen, sulfur, or phosphorus. What molecules give what colors? The whole variety of floral colors are made from four main types of molecules, and modifications of these.

Many red and blue pigments and variations of these colors come from a group of molecules called **anthocyanins** (from the Greek words for flower, *anthos* and for blue, *cyan*). Some anthocyanins are responsible for colors of fruits, red wine, and autumn leaves as well as flowers (we came across them briefly earlier, in connection with the red leaves of Japanese maples). The simplest anthocyanin has fifteen carbon atoms, eleven hydrogens, and one oxygen, with the carbons arranged in rings. Plants are at least as good as animals at building complicated molecules starting from atoms and simple molecules such as water and carbon dioxide.

The pigment molecule that makes flowers like dahlias and snap-dragons yellow is a chemical called an **aurone**, with a structure related to that of anthocyanin. Other yellow and orange flowers, like sunflowers and marigolds, are colored by molecules called **carotinoids**. **Betalains** are the pigments that color flowers such as portulaca and red beets.

Figuring out the structure of complex pigment molecules like anthocyanins required the dedication and skill of fine chemists. To start with, they collected flowers like cornflowers and roses in gardens. Then pure anthocyanin was obtained for analysis by cooking the petals gently in water and using chemical methods of purification. The process is similar to that used for millennia to obtain dyes for cloth. Determining the chemical structure was the first step in understanding the pathway by which flowers construct an anthocyanin. Each step in the pathway is catalyzed by an enzyme, and each of these enzymes is coded for by one of the plant's genes. Plants use the same initial steps to make many anthocyanin pigments. We can get an idea of how many genes are needed to make the molecules, and how efficient plants are in using their resources, from a sketch of the general process.

First, the plant must make the simpler molecules that will be used to construct the anthocyanin. One of these molecules has three carbons and one has nine carbons (both also have hydrogen and oxygen atoms). A different enzyme catalyzes each of the multiple steps the plant uses to make the 3-carbon and 9-carbon starter molecules and each enzyme is made using a different gene. Then a different enzyme combines three of the 3-carbon molecules with one of the 9-carbon molecules to form a larger molecule, **chalcone**. The genes that encode these enzymes would have been switched off in the plant until they were switched on in the petals during flower formation. Although eighteen carbon atoms are in the starting molecules, three are lost in the process of making chalcone. Chalcone has fifteen carbons, five oxygens, and a bunch of hydrogen atoms, and most of these atoms are arranged in rings. Making chalcone is just the beginning, and already a number of genes and enzymes have been engaged. Moreover, behind these are the genes that encode the proteins and RNAs that regulate the genes that code for the enzymes that actually construct the pigments.

Altogether, many genes and the enzymes they encode are involved in producing anthocyanin to color the petals. That's a lot of resources used to attract some insects and birds. Evolution ensured that plants make efficient use of these resources; the same 3-carbon and 9-carbon structures are used to make other important molecules such as fatty acids and resveratrol, the molecule found in red wine that became famous because it extends the life span of mice. This is a common theme in the overall economy of cells; evolution co-opts molecular building blocks for diverse purposes.

Making anthocyanin molecules starting with chalcone requires five or more genes (and the enzymes coded for by those genes), depending on which anthocyanin is required. A cell in a geranium petal uses about seven different enzymes to make its red pigment from chalcone. At least seven different genes are needed, and a mutation in any one of them can mean that the reddish-orange color is altered or lost. Related genes and enzymes as well as others are responsible for making blue and magenta anthocyanin pigments in many different flowers.

The various anthocyanins have different kinds of atoms hanging off the carbons that form rings. The names of these different anthocyanins are long, but some have a reasonable explanation. For example, delphinidins are the molecules that give purple, mauve, or blue color to delphiniums, larkspur, and other blue flowers. To make things easier, we can just call delphinidin the blue anthocyanin and other anthocyanins red (for ones that give red, as in geraniums) and magenta (for the crimson or magenta color of some roses).

Making a yellow aurone molecule also begins with chalcone and requires a separate set of enzymes and thus genes.

Betalains are a different kind of pigment entirely. They are constructed by the plant beginning with an amino acid called tyrosine. This is another example of the efficiency of plant processes. Tyrosine is needed to make almost all of a cell's proteins, and making the amino acid requires a separate set of gene and enzyme; once made, the plant also uses the tyrosine to make betalains. The set of enzymes needed to make a betalain from tyrosine are totally different from that used to make anthocyanins. (Curiously, one of the molecules that is intermediary in betalain construction is also an intermediate, in animals, in the construction of the dopamine neurotransmitter in our brains. Evolution uses the same raw materials for a variety of purposes.) Apparently, flowers that make betalains can't make anthocyanins and vice versa, although no one has figured out why that is so.

Carotinoids have more than thirty carbon atoms and serve a variety of important plant functions beside coloring petals. The orange color of carrots, a root, comes from a carotinoid. Humans use carotinoids to manufacture vitamin A but we are unable to make the carotinoids ourselves. Vitamin A is essential for human health, and that is why eating vegetables like carrots is so important. That is also why scientists have developed engineered rice that contains carotinoids, so-called "yellow rice"; rice is the major staple food for millions of people who lack a good dietary source of vitamin A. Dietary carotinoids are also the molecules that color flamingos pink and the muscles of salmon red. The brilliant red of a cooked lobster shell comes from another carotinoid; before cooking, the shell seems bluish because the carotinoid is tied up with some proteins in cells, but the heat makes the two come apart. The molecular building blocks for these pigments are the same 5-carbon molecules used to make certain plant perfumes, as we shall see in the next chapter.

Not all flowers are just white, red, magenta, blue, or yellow. It seems as though an infinite variety of shades and combinations are possible. How can that be if plants can make only a few different pigment molecules? Genetics works in subtle ways.

A flower's hue depends on more than which dye is present. Monkey flowers (*Mimulus*) are a favorite wild flower of mine, and have been studied by many botanists. There are many different species of monkey flowers and they are commonly found in summer in the western mountains of the USA, especially near streams.

One, called *Mimulus lewisii* (named for Meriweather Lewis of the famous Lewis and Clark expedition), has pink blossoms. The color is from an anthocyanin but there is not very much of it so the flower looks pink rather than red.

Pink flowers are often the result of a low production of anthocyanin pigment. This could be because the switch for turning on one or more of the required genes is missing or not very good at its job. Low pigment production can also be the result of changes in other modulators of gene activity. The result of such changes is that cells may not make as much pigment as they need to produce a rich color. If the color was supposed to be red, the flower may turn out pink instead: if blue, it may turn out sky blue instead of deep blue.

Another way by which a particular pigment can show different shades or colors depends on the acidity or pH of the pigment's environment inside the cell. Pigments are usually stored in small sacs. Sometimes, these sacs have a different level of acidity than the cells that contain them. Depending on just how acidic the sac is, some pigments may turn different colors. Gardeners know that hydrangea bushes produce pink or blue blossoms, depending on the pH of the soil. (Hydrangeas, by the way, do not have petals; what look like petals are actually sepals.) In acid soil, the flowers are colored blue by a delphinidin dye (as long as there is a bit of aluminum in the soil). Adding lime to the soil decreases the acidity (increases the pH) and the same bush shows pink blooms. No changes in genes are required; the pigment is still delphinidin. Color variation can be quite complex because the acidity of the sacs is itself determined by a set of genes. A mutation in such a gene could change the color of a flower without a gardener's intervention or a change in the genes that made the pigment. This occurs, for example, in petunias, where certain mutations change the acidity of the sacs, causing red petunias to become bluish.

Even the shape of the petal's cells influences the color because it can change the way the light is reflected back to our eyes. And the way the molecules of pigment are packed together in the sac can also influence the color we can see.

Color is even more complex and subtle if the flower can make more than one kind of pigment. *Mimulus lewisii* makes carotinoids as well as anthocyanin but the yellow carotinoid color only occurs as two yellow spots on the pink blossom (Plate 8). The yellow spots serve as guides for pollinating bees. Another kind of monkey flower, *Mimulus cardinalis*, is a particular shade of red. It makes more anthocyanin than its cousin so the hue is darker, and it also makes carotinoid but it makes that pigment throughout the petal rather than in two spots. The combination of the two pigments gives M. *cardinalis* its special color. But bees don't see this color well and are not attracted. Instead, M. *cardinalis* is pollinated by hummingbirds (Plate 3). The blossom has also evolved to be narrower than that of M. *lewisii*, and this accommodates the hummingbirds' narrow beaks. All of these properties of the two cousins depend on genes—the genes that are needed to make the colors and shapes of the flowers, and other genes that can control where (in what cells) in the petal the color is made.

The story of M. *cardinalis* illustrates the fact that some flowers produce several different pigments simultaneously, for example both red and yellow. Such flowers, like M. *cardinalis*, commonly look more red than yellow. But the flower will be yellow if a mutation occurs in one of the genes needed to make the red pigment. And the offspring of that plant will also produce yellow flowers, so long as the carpels receive pollen from the same flower or a close relative. These kinds of changes are reasonably common in nature. Plant breeders depended on such random events to grow plants with uniquely colored flowers that are treasured by gardeners and nursery people. Today, plant breeders combine their traditional methods with genetic manipulation to do the same thing more precisely and quickly.

Some years ago, I discovered a wonderful plant, a native of South America, for sale in my local nursery. Torenia produces abundant purple flowers in hanging pots even in the shade, and doesn't seem to mind our very hot summers in DC. Now, genetic engineering has produced torenia plants with yellow blossoms. The Japanese scientists who did this had some fancy engineering to do. They had to stop the synthesis of the purple pigment, and they did this by introducing a gene for a small RNA molecule that interferes with production of an enzyme required for normal anthocyanin synthesis. (Note that a small RNA molecule can change the way a gene does its job without changing the gene itself.) They also introduced two genes that provide enzymes needed to make aurone from chalcone. In the spring of 2007, I found the yellow torenia in the local nursery. It was, of course, more expensive than the common, purple variety, but it did very well in pots in my garden.

Many flowers are a single color or shade, but others have amazing displays of sectors, spots, or streaks of different colors. It's not surprising that there are various ways to form these multicolor patterns.

Red Star petunias are well known to gardeners. The vivid red blossoms sport a white star emerging from the center and extending into the red petals. The markings on this popular petunia variety reflect the pattern of activity of one gene important in the path to making chalcone. The white areas have much less of that gene's messenger RNA than is present in the red regions. It is likely that the gene is expressed equally well in both areas of the blossom but that the messenger RNA is much less stable in the white regions. It's also likely that the messenger is destroyed by the action of a small RNA. While that conclusion is somewhat tentative, it makes a lot of sense. But it doesn't explain why the white areas form the star pattern as opposed to affecting the entire blossom. One likely explanation emerges from observing that the white star forms around the veins in the blossom. If a destructive RNA is spread in the floral sap, it would reach the regions around the veins before spreading into the rest of the flower.

A late-blooming azalea bush in my garden has white blossoms. Soon after we planted it forty years ago, I was surprised to see that some of the blossoms had randomly placed pink sectors or streaks. Each spring, the same thing happens on this particular plant. It's not a rare phenomenon; it occurs also in other plants like snapdragons, petunias, and morning glories. Darwin studied a variety of snapdragon that had white blossoms marked by red stripes and spots. He knew nothing about Mendel's experiments or genes and could not even speculate on how such streaks of color might arise. Today, there is a plausible story that can explain why colored areas sometimes occur on the white background of the azalea petals.

Keep in mind that each cell has its own DNA and genes and makes its own enzymes; that DNA and those genes are like the ones in all the cells of that plant and the same as those in the seed that produced the plant. During the life of a cell, its DNA and genes can occasionally mutate, and the change is passed on to all the daughter cells during cell division, without having any effect on other cells in the vicinity.

A whole set of enzymes and the corresponding genes must work properly to make the red anthocyanin pigment of the azalea blossoms on my bush. Imagine that a debilitating mutation occurred in one of the genes the plant needs to make the red anthocyanin. Without that gene and its enzyme, no anthocyanin can be made and the bush produces white flowers. Imagine further that the mutation, the change in the DNA, is an unusual one. It is the result of a random event, the disruption of the gene by insertion of a jumping gene into its DNA; the jumping gene originated from elsewhere in the cell's DNA. The result is that the disrupted gene no longer functions and the anthocyanin is not made.

One of the interesting things about jumping genes is that they can, occasionally, jump out, and if so the original DNA sequence is restored. Imagine now that in one cell of a developing white petal, the jumping gene jumps out. That cell will now be able to make red pigment. As the flower grows bigger, the cells in the petal multiply. Most of them will still be white. But all of the cells that are made by divisions of that red cell, its daughter cells, will also be able to make red pigment. If there are enough of them, they will be visible as a red spot or a red streak on the otherwise white flower (Plate 9).

It's hard to study genes in azaleas because the bushes are slow growing and often propagated by cuttings not seeds. But the same explanation of the jumping genes works for several annuals such as petunias and snapdragons. There are tricks that can make jumping genes jump more often than they do in nature. From the seeds of a few plants, many different mutant plants grow and those with mutations in flower color are easily identified and then bred for study. For example, a mutant petunia with white streaks on its red flowers was picked out from among such plants. Taking advantage of the presence of the jumping gene DNA, the DNA of the affected gene could be purified. That DNA was then used to purify the original, unmutated gene. A jumping gene's DNA provides a kind of a tag for identifying and purifying normal genes. The identified normal genes can lead to identification of the proteins or RNAs they produce.

The most prized, and most expensive varieties of tulips during the seventeenth-century European craze were the so-called "broken" tulips. They had feathery petals and bright flames of red and other intense and unusual color schemes, like the modern variety called Snow Flame. Viruses were responsible for the unusual color schemes. The virus' DNA, like jumping genes, can insert into the DNA of the plant. The daughter cells of the originally infected cells produce the unusual color markings.

Color and color patterns are among the properties that attract insects, birds, and ourselves to flowers. It seems that people have watched for and experimented with unusual flower color since prehistoric times. Today, there is a great deal of interest in the colors of orchids. Once rare and always prized, orchids are now relatively common and many people grow the plants and experiment with production of new varieties. People all over the world breed orchids to try to obtain distinct coloration and shape, and they compete at huge international orchid shows. The three petal-like sepals and the three petals vary greatly in shape and size, including some with ruffled petal edges. The variability of color is even more striking in both the range of color and the versatility of contrasting markings. Still, orchid breeders and the general public are aware of only a few of the more than 26,000 orchid species known, many of which only occur in the tropics.

Flower color, evolved to attract pollinators, has become a magnet for gardeners and plant breeders. That other adaptation of flowers that draws us—their scent—forms the topic of the final chapter.

Chapter 12

THE PERFUME FACTORY

In mid-June, a glorious perfume fills the Piazza del Duomo in the Italian hill town of Ravello. The whole town looks glorious too, but that is a different subject. The perfume is strong enough to swamp out the smell of exhaust from the local bus that labors up the mountain from the Mediterranean coast and pulls up twenty-five yards from the piazza itself. My immediate thought was that it must be jasmine (*Jasminum officinale*). But it's another plant with a similar leaf, flower, and scent, *Rhyncospermum jasminoides*.

This memory tells us something about plant perfumes. The scents produced by plant petals diffuse into the air. Such rapidly evaporating substances are called volatiles (from the Latin *volare*, to fly). They escape a plant even at moderate temperatures. Water, in contrast, takes a lot more energy to evaporate, and only does so at a significant rate as steam at its normal boiling point of 100°C (212°F). Volatile molecules from plants are generally quite small, smaller than the color pigment molecules.

The perfumes of flowers are made in the cells of the petals. Those fragile petals are amazing chemical factories. Other parts of plants, such as leaves and the bark of some trees, also make scents. The smell of a pine forest is an unforgettable experience, although pines are not flowering plants. Many plant odors, like that of mint, are easily detected by rubbing a leaf or stem. And not all plant volatiles smell pleasant. The leaves and stems of the Sky Pilot (*Polemonium viscosum*), which grows in the Rocky Mountains, produce a strong, skunk-like odor that is easily smelled several yards away. And the rarely-blooming Titan Arum (*Amorphophallus titanum*) has the stench of rotting flesh, drawing the flies and beetles needed for pollination, not to mention curiosity seekers to botanical gardens.

Flower scents, like the colors and shapes of petals, are there to attract pollinating insects and birds, and ward off predators. Some volatiles attract enemies of plant predators in a natural demonstration of the ancient adage "the enemy of my enemy is my friend." The scents evolved in concert with friends and foes alike to ensure seed formation and plant reproduction. Scent production is often timed to be highest in the afternoon, when insects are out in force, to optimize the chances for pollination. As with flower colors, the appeal of many floral scents to humans is a fortunate byproduct; we were not even around when they appeared. In fact, our purposeful breeding and selection to produce showy flowers often results in a loss of perfume.

Some volatiles evolved to dissuade predator insects, and protect the plants from attacking fungi and bacteria. Such natural defenses are part of the reason why spices have been big business for millennia. Many of them, like cloves, come from South and East Asia, and their importation into Europe spurred trade for centuries. Voyages of discovery, including that which led to Columbus' first encounter with America in 1492, were, after all, funded in the hopes, initially, of improving access to eastern spices, and subsequently for discovering the resources, including the new plants and spices, of the vast continent. Some of the earliest explorers of the Western Hemisphere were botanists.

Oils derived from flowers have been used for centuries, but it was from the nineteenth century onward that the large-scale manufacture of plant volatiles as well as dyes was established. Perfumes extracted from flowers are costly and they have uses beside personal adornment. The flavor of food is a combination of what taste buds perceive and the smells that the foods emit when chewed. But before the perfume and food flavoring industries could grow into large enterprises, they had to hire chemists to find out what the volatile molecules were, and to learn to synthesize them in the laboratory. A good deal of information about scented volatiles was collected by these industries. The molecules could easily be extracted from collected flower petals and other plant parts by soaking them in alcohols. Even today there is a cottage industry of scents obtained in this way. Volatiles for chemical analysis are usually collected by enclosing flowers in contraptions that don't allow the gases to escape. They are then analyzed by techniques such as mass spectrometry.

For all the effort, commercial perfumes rarely smell like flowers. Expensive, fancy bottles labeled jasmine or gardenia may smell wonderful but they are sad substitutes for the real thing. One reason is that flowers generally produce very large mixtures of different volatile molecules, as many as a thousand. Some of these fall into related chemical groups and although they differ very slightly in chemical structure, they can produce very different smells. In closely related flowers, the volatile molecules can vary both in relative amounts (reflecting differential regulation of the genes and gene products needed) and in their chemical structures (reflecting the activity of genes evolved to produce the enzymes needed for synthesis). It's not easy to figure out which components of a mixture are important for attracting insects or birds or for achieving a perfume attractive to people. It is especially challenging because our own sense of smell depends on a complex set of nerve cells and often differs from one person to another. The manufacture of the odors depends on a plant's genes, and the ability of animals, including ourselves, to smell those odors depends on the animals' genes.

Today, many plants once known for their perfume barely retain a smell at all. Roses are a prime example. Breeding is ultimately about selecting genes for the desired traits, even though it began long before genes were recognized. Many rose varieties were bred to enhance qualities like color, longevity, and, for cut flowers, the ability to travel long distances to meet the huge demands of flower lovers around the world. In the process, perfume disappeared. Those that have retained their perfume, such as tea roses, are now much prized.

As with colors, the chemistry of volatiles depends on the presence of genes that encode protein enzymes. These enzymes act in sequence to produce the complex scented molecules from precursor molecules whose presence depends on still other genes and enzymes. The relative amounts of the different molecules depend in turn on other genes that code for RNAs and proteins important for the regulation and modulation of the genes required to manufacture the scents.

When we smell a rose, we are picking up a mixture of several hundred different molecules. Each one of these is the result of a series of genes and the enzymes they code for that enable particular chemical reactions in the rose petals. Many of the volatile molecules are made from the amino acid **phenylalanine**.

All living things and their cells need phenylalanine because it is an essential component of many proteins. Plants have no problem making phenylalanine. But many animals, including ourselves, are unable to make this essential molecule. We rely on our diets to provide phenylalanine. Nevertheless, an inherited disease, phenyl-ketonuria, is caused in susceptible individuals by excess phenylalanine. The excess of the amino acid causes the mental deficiencies associated with the disease. Newborn infants who test positive for the mutation causing phenylketonuria are fed diets low in phenylalanine.

Plants manufacture phenylalanine from simpler molecules, via a set of genes that code for the necessary protein enzymes. Phenylalanine is a close relative of tyrosine, the amino acid used by plants to manufacture the betalaine pigments, and it too is an "aromatic compound," with a ring of carbon atoms. The difference in chemical structure between the two is simply that tyrosine has an additional oxygen (in the form of an –OH group attached to the carbon ring). In fact, mammals make tyrosine from phenylalanine (plants use another path). The list of pleasant-smelling molecules derived from phenylalanine and tyrosine is long.

Plants make phenyalanine and tyrosine so that they can make proteins. But evolution, being opportunistic, makes use of the amino acids for other purposes too. Each use depends on evolving one or more additional genes that code for the enzymes that make the aromatics as well as the proteins and RNA needed to make sure that the genes are turned on in petals at the right time. A number of the aromatic volatiles will have originated from gene duplication events followed by mutation of the copies, a pattern we have now met several times. It is one of the most powerful ways in which variations become available on which natural selection can act.

To make a volatile aromatic from either of the amino acids phenylalanine or tyrosine requires chemical surgery on the amino acid by one or more reactions catalyzed by particular enzymes. One such reaction removes the amino group ($-NH_2$) from the amino acid. If the starting molecule is phenylalanine, the result is a molecule called cinnamic acid; if the starting molecule is tyrosine, the result is coumaric acid. The only difference between cinnamic and coumaric acids is that coumaric acid has the same additional oxygen atom in the form of an $-OH$ group as does tyrosine. Most, but not all, plant aromatics start out as one of these two molecules.

The name cinnamic acid shouldn't be a mystery. It is what gives cinnamon its familiar smell. Cinnamon is the dried bark of certain evergreen trees of the genus *Cinnamomum* in the laurel family, which reminds us that many plant parts beside petals make aromatics. The enzyme that carries out the removal of the amino group from phenylalanine to produce the acid is called PAL, encoded by the gene *PAL*. Most plants have more than one *PAL* gene. Arabidopsis, for example, has four *PAL* genes, and these are active to different extents in different parts of the plant. It makes some sense to have several *PAL* genes because that same phenylalanine minus its amino group, as cinnamic acid, gives rise to many plant molecules beside volatiles. Among those molecules are lignin, the huge molecule found in tree wood, and the flavonoid pigments used in coloring flowers. Some plants use PAL to start the long series of reactions leading to chalcone, the molecule that is eventually converted to the anthocyanin colors.

Eugenol is one of the plant aromatics made from coumaric acid and thus, indirectly, from tyrosine. Many of us encounter eugenol in the dentist's surgery, where oil of cloves is often used as a mild anaesthetic. Cinnamon and basil too depend, in part, on eugenol for their spicy, herbal smells, and the frankincense and myrrh mentioned in the Bible owe their medicinal qualities to eugenol.

Another pathway to aromatics from phenylalanine involves two excisions on the amino acid. Here, both the amino ($-NH_2$) and the acid ($-COOH$) groups which characterize it as an amino acid are removed. The resulting molecule is the starting point for making many other aromatic molecules. The level of the enzymes required to carry out this surgery in rose petals is most abundant in mature flowers, and in the late afternoon, when it is important to attract pollinating insects. This is another example of the working of the biological clock described in Chapter 5. Evolution has ensured that genes are most active when they are needed.

The identification of the gene responsible for the enzyme that removes the acid group from phenylalanine required real detective work. Plant genome data banks were searched for sequences that, by analogy with genes known in other organisms, might produce an enzyme that removes the acid group from phenylalanine. Scientists hit the jackpot when they found plant DNA sequences similar to the sequence of an animal gene that removes the acid group from a molecule called dopa that is related to phenylalanine. Familiar? That's the same dopa that is used as a treatment for Parkinson's Disease. This DNA segment was most active in plants at the times and in the flower parts, petals and ovaries when production of the volatile molecules from phenylalanine was highest. When activity of the gene in petunias was turned down experimentally in mutant plants, production of the aromatic stopped. The same was true for the rose version of the gene.

The petunia and rose forms of this gene code for protein enzymes that are about sixty-five percent the same as animal enzymes that remove the acid portion from dopa and are similar to other plant enzymes that also remove the acid portion from other molecules. Together, all these genes belong to a family of related genes. It makes sense to conclude that they all evolved from some common ancestral gene.

Flowering plants have many more genes that code for enzymes required for production of other aromatics. Where did they all come from? Probably most, if not all of them, are related to genes that are important for other plant functions, and arose from past gene duplication events. That is what appears to have happened during the evolution of the genes responsible for the fragrant "tea" smell characteristic of popular tea roses. When ancient breeds of Chinese roses made their way to Europe late in the eighteenth century, it was recognized that they had a different perfume from European roses. Many years later, these unique smells were associated with particular compounds. By then, hybrids between Chinese and European roses had been bred. The hybrids, known as tea roses, are especially popular and one reason is their strong and appealing perfume, an inheritance from the Chinese parent of the hybrid. Among these perfumes one aromatic molecule (3,5-dimethoxytoluene, abbreviated to DMT) can contribute as much as ninety percent of the volatiles produced by the flowers. European rose petals don't produce much, if any, of this molecule.

The DMT molecule is related to other plant aromatics constructed on a core ring of six carbon atoms, a few of which are decorated with assortments of carbon, hydrogen, and oxygen atoms. Various genes and enzymes give the plant the ability to make such decorated rings. Two enzymes coded for in Chinese rose genomes and active in Chinese rose petals can make the particular modifications leading to DMT. Why can't the European roses do this? Because they don't have the set of genes needed to make the proper modification. Two very closely related but distinct genes lead to the appropriate chemical changes in roses with a Chinese rose heritage somewhere in their past; they are called *OOMT1* and *OOMT2*. Roses of purely European origins have only one of the two genes, while both proteins are needed to modify the aromatic ring in the right way to produce DMT. The 350 amino acids in the two enzymes OOMT1 and OOMT2 are ninety-six percent identical, and a change in just one amino acid out of the 350 is likely to be responsible for the difference in what they can do in the cells of petals. All this suggests that there was, initially, a single *OOMT* gene that became duplicated, and that one of the two copies then acquired mutations in its DNA and, as a consequence, changes to the amino acids of the protein enzyme for which it codes.

Which gene came first? If the *OOMT* genes in many different roses are compared, most have *OOMT2* but only rose varieties with Chinese rose ancestors have *OOMT1*. The evolutionary tree of roses has features that make it likely that the Chinese roses appeared later in time than other roses. If so, that would be a strong clue that *OOMT2* has been around for a longer time than *OOMT1* and that it was *OOMT2* that became duplicated.

Producing roses with smells that please people could not have been the reason for the success of this gene duplication and change by mutation. Why then did the new gene survive and succeed? It turns out that bees, important pollinators of roses, seem to sense DMT. Perhaps the extra dose of DMT gave the Chinese roses a reproductive advantage in nature.

There are many plant volatiles that have no relation to phenylalanine or tyrosine. Some of the most important are related instead to turpentine. The word turpentine does not bring flowers or wonderful perfumes to mind. Rather, it makes us think about paint. But turpentine itself comes from the bark and wood of evergreen trees, mostly pines. And it gave its name to a bevy of other compounds called terpenes that contribute to the perfumes of many flowers. (Why terpenes and not turpenes? In German, turpentine is *terpentin* and German chemists were the first to study the chemicals.)

Turpentine is a mixture of terpenes. What all terpenes have in common is that they are constructed by joining molecules that contain five carbons. Some terpenes have ten carbons, some fifteen, and so on through multiples of five. The base unit molecule with five carbons, isoprene, is made in several plant parts. It is volatile and much of it is released into the air: more than half a billion tons every year. It is one of the molecules involved in the synthesis of ozone in the troposphere. Terpenes with ten and fifteen carbons are important to the smell of flowers, perfumes, herbs, and spices. Terpenes with fifteen carbons serve as antibiotics for plants and also help to ward off herbivores. Even the production of vitamin A involves molecules partly built from isoprene units. And more than 100 isoprene units combine to make rubber. Plants have at least two pathways by which isoprene can be made. Each depends on several enzymes and the genes that encode them.

Plants are not alone in making terpenes. Cholesterol is made by animals, including ourselves, from the same 15-carbon terpene that plants use to make perfumes. Not only that, the genes and proteins that make the 15-carbon terpene in plants and animals are similar enough to indicate that they had the same parent genes way back in evolution. Plants, however, don't make much cholesterol at all.

The gene and enzyme responsible for combining two 5-carbon isoprene units catalyzes the first step in the manufacture of many terpenes. The 10-carbon product is called geranyl diphosphate, and the enzymes that do the joining in different plants are closely related, reminding us that all flowering plants have, way back in the history of the planet, a common ancestor.

Scientists interested in floral scent have made "gene libraries" containing thousands of different DNA sequences, each one representing a gene that is switched on in petals. Each of the different petal DNA segments was linked to other DNA sequences that can reproduce many times over in bacterial cells. This "recombinant" DNA library was then inserted into bacteria so that each bacterial cell received only one of the recombinants. The bacteria were then spread out on a plate containing nutrients and allowed to multiply. The plate then contains small colonies of cells, each colony containing the offspring of the initially spread cells. Each bacterial colony then contains a unique DNA sequence originally from the petal DNA. To be useful, the sequence of the As, Ts, Gs, and Cs in the inserted DNA segment in each colony of cells in the collection needs to be determined and entered into an accessible data base. This provides a unique identity for each cell colony, an identity defined by the original segment of petal DNA. That's the first step.

The second step is to search among the many different petal DNA segments for those that encode proteins similar to known genes. In this example, a gene encoding an enzyme that makes geranyl diphosphate in peppermint plants (*Menthus piperata*) was already known. The task then was to look in a snapdragon DNA library for genes (i.e., DNA sequences) similar to the one in peppermint plants. When that was accomplished, those bacteria were allowed to make the protein coded for by the similar snapdragon gene. Sure enough, that protein did indeed catalyze the formation of the terpene geranyl diphosphate from 5-carbon molecules. This general procedure underlies the isolation of many of the genes involved in making flowers.

Like many of the activities in plants and flowers, the gene for the enzyme that catalyzes the formation of geranyl diphosphate (the enzyme is known as geranyl diphosphate synthase) is active in petals at particular times. The timing of gene expression can be assessed by using the DNA from the recombinant as a probe to detect when and how much of the messenger RNA is produced. Sure enough, the amount of messenger RNA and protein in the petals rises when the flower bud begins to open, and it is found mainly in the petals, with very little in sepals, carpels, or stamens. It is by using techniques like these that scientists are piecing together the genetics of floral scents.

What about the marvelous perfume of jasmine whose memory was evoked at the start of this chapter? There are more than forty varieties of jasmine, and each has a slightly different smell. All these perfumes are made of varying amounts of a mixture of as many as 300 different molecules. No wonder the perfume is so hard to duplicate. Some of these molecules are aromatics like eugenol, and others are terpene derivatives. But one important component is another sort of molecule entirely, and its name itself comes from jasmine: methyl jasmonate. The making of methyl jasmonate is a further example of how flowers utilize molecules that are made in all plant cells for a special purpose.

One of the fats in plants is a molecule called linolenic acid, with a long string of eighteen carbon atoms. The name comes from flax plants, the source of linen. There are many species of flax plants and they all are part of the genus *Linum*. Another common substance associated with flax, which has industrial and even gastronomic uses, is linseed oil, which comes from the seeds of flax plants. Even the word linoleum has the same root because that material was made, originally, from linseed oil. Linolenic acid is one of those molecules popularly known as an omega fatty acid. Omega fatty acids are essential for normal growth and development, but neither we nor other mammals can make them for ourselves. We don't have the genes necessary to make the enzymes required to produce linolenic acid. We depend on foods such as certain fish, beans, and nuts in our diet to supply omega fatty acids.

Plant cells do have genes that enable them to make linolenic acid. And evolution took advantage of its presence to develop a variety of other useful molecules. The carbon atom at one end of the long, 18-carbon string in linolenic acid is part of the acid (–COOH) group of the molecule. That acid group will still be there when, after several chemical steps, jasmonic acid is produced from linolenic acid. Jasmonic acid has only twelve carbons, six fewer carbons than linolenic acid. Each of the chemical steps involved in its production depends on a protein enzyme and the gene that codes for the enzyme.

Jasmonic acid does a lot of important things for the plant. One is to adjust the activity of various genes after an insect wounds the plant. Jasmonic acid is also important for pollen production; without jasmonic acid, the flowers are sterile unless pollinated by a plant that does make jasmonic acid. But jasmonic acid has no smell. For scent, two related molecules must be made. The flower needs a gene for a protein enzyme that will remove the acid group to make jasmone, with eleven carbons. Another gene and its enzyme add a methyl group (–CH_3) to form methyl jasmonate. Both jasmone and methyl jasmonate are volatiles that contribute to the perfume of jasmine and other plants. Perfume makers add such compounds to their concoctions.

The colors, shapes, and perfumes of flowers have all evolved to attract particular pollinators to the different plant species. But many flowers provide an additional attraction: nectar. Feasting on nectar is a reward for the pollinators' services. Another of the many things that jasmonic acid does for the plant is to stimulate the production of nectar. Evolution does not waste resources; plants that are pollinated by wind-born pollen do not usually make nectar.

The ancient myths report that the gods could live on nectar alone. No wonder. It contains sugars, amino acids, and proteins so it provides a pretty complete diet. Beside being tasty, many nectars also include attractive fragrances, antibiotics, and molecules that deter unwanted animal visitors. The manufacture of all these components, their collection together in one substance, and their release from cells is an important part of flower construction.

Although the production of nectar is surely as important to the whole system as the production of color and smell, plant biologists are only just beginning to illuminate the processes involved. The actual composition of the nectar varies between flowers of different species and even between individuals in one species. Each kind of flower makes a different mixture that seems to have evolved in coordination with the particular animals that visit that plant or microbes that can infect it and compete for the nutrients in the nectar. Bees, for example, seem to be attracted to caffeine, and that molecule occurs in some nectars. Even when the same components are in the nectar, their relative amounts can vary from one kind of plant to another. Plants pollinated by ants and butterflies, for example, have more amino acids than those visited by birds and bats; the latter animals can satisfy their need for amino acids in other ways.

By far the most abundant component of nectar is sugar. Some of the proteins in nectar appear to be enzymes that can incapacitate or destroy pathogenic invaders such as yeasts and other microbes. Other enzymes convert the sugars to different forms that are preferable to the taste of pollinators. The enzyme invertase is often found in nectar; it converts more complex sugars into glucose, which is favored by pollinators. A good deal is known about how plants manufacture the various sugars typically found in nectar. One protein, called SWEET9, accounts for the release of one particular sugar from cells and the deposit of that sugar in nectar. But less is known about the production of the other nectar components.

Nectar often begins to accumulate in a flower just as the petals and sepals emerge from the flower bud. In some flowers, it's possible to identify a special group of cells, known as a nectary, that produce nectar. In others, the cells that produce nectar seem to be spread thinly around the surface of the petals and some plants produce nectar in other parts of the plant beside flowers. Once the nectar is complete, it needs to be released from cells. Jasmonic acid—the same substance that is used to make some floral scents—stimulates nectar secretion in at least several species of plant, if not all.

A FINAL WORD

Plants invest a lot of their available resources in the manufacture of color, perfume, and nectar for their flowers. All this is to ensure that they draw pollinators. Flower shapes have often coevolved with pollinators to ease their access to nectar and to the pollen, increasing the efficiency of pollination. Plants may be unable to move themselves, but their blossoms draw in the animals that will help them to fertilize their eggs and spread their seeds.

The amazing variety of the floral world is the consequence of evolutionary changes that promote the effective reproduction of flowering plants. We humans are the accidental beneficiaries. Our lives are enhanced by flowers, their vivid colors, and intoxicating perfumes. Little wonder that, across the world, flowers have inspired art and culture over the centuries. Even more important, we and other animals depend, for our food, on the successful pollination of plants which results in the production of seeds, giving us seed grains like maize and wheat, and a whole variety of vegetables and fruit.

And behind all the riot of color, shape, and scent of the floral world lie all the genes, and the proteins and RNAs they code for, including genes that regulate the activity of other genes that code for the making of the parts of a flower, or enzymes that produce pigments and perfumes. The activities of all these genes are modulated by a variety of epigenetic marks that are still being elucidated. This book provides only a glimpse of the complexity of the genetics involved in flowering. There is much we do not yet understand. It is a story we are still just piecing together, a story that is growing and unfolding even as I write.

GLOSSARY

amino acids chemical building block of *proteins*

anther top of the *stamen*, where pollen is produced

anthocyanin plant pigments that color flowers blue, purple, and red hues

Arabidopsis *genus* of a plant that is used in much of contemporary plant research

aurone complex molecule made by plants that produces petal color, especially yellow hues

auxin plant hormone that stimulates growth

base (in DNA) any of the four molecular units in DNA—adenine, thymine, cytosine, and guanine (labelled A, T, C, and G, respectively)

betalain certain red and yellow plant pigments

carotinoids natural plant pigments giving red and yellow hues

carpel female reproductive element in a flower; one or more carpels make up the *pistil*

chalcone complex molecule made by plants that contributes to fragrance

chlorophyll molecule in plants that collects light energy from the Sun, and gives rise to their green color

chromatin combination of DNA and protein usually found in the nucleus of plant cells

chromosome individual package of DNA and *protein* contributing to chromatin

codon group of three DNA *bases*, or their RNA equivalent, that specifies an *amino acid*, e.g., AUG for the amino acid methionine

disc flowers many tiny flowers at the center of a compound flower and surrounded by *petals*

DNA large molecule that holds genetic information

epigenetic marks extra chemical groups (such as *methyl groups*), affixed to any of the four common DNA *bases*, which can influence the effects of *genes*

epigenetic writer enzyme that adds the *epigenetic marker* group to a DNA base

eigenetic reader a protein that recognizes and acts on the presence of the *epigenetic mark*

epigenetic eraser enzyme that removes one or more *epigenetic marks*

florigen a plant hormone that initiates flower formation

gamete egg or sperm, the reproductive cells of sexually reproducing organisms

gene basic unit of inherited information—a segment of a long DNA (and occasionally an RNA) chain

protein gene DNA segment containing the *codons* (the *genetic code*) that specify the sequence of *amino acids* for a particular protein

RNA gene DNA segment specifying the kind and sequence of RNA *bases* for an RNA molecule (e.g., ribosomal RNA)

genetic code information embodied in various triplets of DNA *bases* that directs the incorporation of particular *amino acids* into proteins

genome totality of DNA and thus of genetic information in the cells of any organism

genus a category of related but not identical organisms. For example, the genus *Chrysanthemum* includes what are commonly called daisies as well as flowers called chrysanthemums

gibberellin plant hormone that regulates plant growth and development

histone *proteins* that interact with DNA in forming *chromatin*

intron noncoding segment interrupting the coding region(s) of a *gene*. Introns are spliced out of the functioning RNA copy of a gene

meristem group of growing, not yet differentiated plant cells, usually at the tips of shoots and roots

methyl group carbon atom linked to three hydrogen atoms; methyl groups can act as *epigenetic marks*

molecular cloning isolation, copying, and multiplication of a particular DNA segment from a genome

mutation change in genetic information encoded in DNA as a consequence of a change in the DNA sequence

nucleosome package consisting of a DNA segment (about 140 base pairs in length) wound around a group of *histones*; the nucleosome is a subunit of *chromatin*

petal part of a flower that is flat like a leaf and usually colored other than green

phenylalanine one of the twenty essential *amino acids* that are the building blocks of proteins

photosynthesis process by which green plants convert the Sun's energy into chemical energy to construct the plants' components

phytochromes plant pigments that absorb red light

pistil the organ of a flower that contains the egg, often described as the female organ of a flower; it may be composed of one *carpel* or several fused carpels

proteins large complex molecules constructed of many *amino acids*

ray flower long floral elements surrounding the central disk of composite flowers such as sunflowers, in which they are commonly referred to as *petals*

RNA ribose nucleic acid that carries the information encoded in DNA to the places in the cell where it is used

messenger RNA RNA that is copied from the DNA of each gene provides the information to the protein-synthesizing components of cells, the ribosomes

sepals usually green leaf-like parts that cover the flower bud and are then typically found beneath the petals when the flower opens

species group of related individuals (plants and/or animals) that can breed with one another to produce viable offspring

stamen organ of a flower that contains the pollen; often described as the male reproductive organ

stem cell cells found in animals and plants that can, through cell division and differentiation, give rise to many different kinds of cells

stigma the tip of the *carpel* or *pistil*, where pollen collects

style the shaft of the *carpel* or *pistil*

FURTHER READING

Bangham, J. A new take on flower arranging. *Nature Review Genetics* 6 (2005), 2.

Benfey, P. N., and T. Mitchell-Olds. From genotype to phenotype: systems biology meets natural variation. *Science* 320 (2008), 495–7.

Buchanan, B. B., W. Gruissem, and R. L. Jones. *Biochemistry and Molecular Biology of Plants*. (Rockville, MD: American Society of Plant Biologists, 2000).

Chamovitz, D. *What a Plant Knows*. (New York: Scientific American/Farrar, Straus and Giroux, 2012).

Grant, V. *The Genetics of Flowering Plants*. (New York: Columbia University Press, 1975).

Jack, T. Molecular and genetic mechanisms of floral control. *The Plant Cell* 16(Suppl) (2004), S1–17.

Levy, Y. Y., and C. Dean. The transition to flowering. *The Plant Cell* 10 (1998), 1973–89.

Pavord, A. *The Naming of Names*. (New York: Bloomsbury, 2008).

Reddy, G. V., M. G. Heisler, D. W. Ehrhardt, and E. M. Meyerowitz. Real-time lineage analysis reveals oriented cell divisions associate with morphogenesis at the shoot apex of *Arabidopsis thaliana*. *Development* 131 (2004), 4225–37.

Science Magazine. 320/5875 (25 April 2008). Special issue on plant genomes.

Singer, M., and P. Berg. *Genes and Genomes* (Mill Valley, CA: University Science Books, 1991).

Smyth, D. R., J. L. Bowman, and E. M. Meyerowitz. Early development in Arabidopsis. *The Plant Cell* 2 (1990), 755–67.

Swiezewski, S., F. Liu, A. Magusin, and C. Dean. Cold-induced silencing by long antisense transcripts of an *Arabidopsis* Polycomb target. *Nature* 462 (10 December 2009), 799.

Wigge, P. A., M. C. Kim, K. E. Jaeger, W. Busch, and D. Weigel. On the origin of flowering plants. *Science* 324 (2009), 28–31.

PUBLISHER ACKNOWLEDGEMENTS

We are grateful for permission to include the following copyright material in this book.

Excerpt from THE COLOUR PURPLE by Alice Walker. Copyright © 1982 by Alice Walker. Used by permission of Houghton Mifflin Harcourt Publishing Company. All rights reserved.

The publisher and author have made every effort to trace and contact all copyright holders before publication. If notified, the publisher will be pleased to rectify any errors or omissions at the earliest opportunity.

PICTURE CREDITS

INDEX